Data of the Planetary System

Data of the Planetary System

edited by THOMAS ROATSCH

AKADEMIE-VERLAG BERLIN 1988

Authors:
Dr. Th. Roatsch
Dipl.-Phys. M. Danz
Dipl.-Phys. G. Elter
Dipl.-Phys. K.-H. Fornacon
Dr. D. Hartke
Dr. E. Kuehrt
Dipl.-Geol. J. Matthes
Dipl.-Ing. C. Roeser
Dr. U. Schmit
Dipl.-Min. H. Schulze
Dipl.-Phys. F. Spahn

Institut für Kosmosforschung der Akademie der Wissenschaften der DDR, Berlin

ISBN 3-05-500426-4

Erschienen im Akademie-Verlag Berlin, DDR-1086 Berlin, Leipziger Straße 3—4

Lizenznummer: 202 · 100/443/88
Printed in the German Democratic Republic
Gesamtherstellung: VEB Druckerei „Thomas Müntzer", 5820 Bad Langensalza
Umschlaggestaltung: Wolfgang Janisch
Lektor: Dipl.-Met. Heide Deutscher
LSV 1497
Bestellnummer: 763 774 6 (9089)
01800

CONTENTS

0. INTRODUCTION

The study of the solar system during the last twenty years has experienced a dramatic growth in the availability of observational data and a continued growth in the level of scientific interest. The interest in solar system data is presently high because of recent space probe data returned of the outer planets and comets.

"Data of the planetary system" contains the characteristic data of all planets with their satellites and rings, the asteorids with orbits between Venus and Mars, which are important for future space missions, and the latest observations of all comets and the properties of their environment. Furthermore, the important data of the interplanetary plasma and the mineralogical components of the meteorites found on the Earth are involved. The plasma quantities used in this book are explained in a formulary (part 3.1.).

The references are given in brackets and listed at the end of the book whereas the missions, which measured the plasma data, are given in fraction strokes below the data and listed in part 3.8.

Because of the inevitable conversion to the inernational system of units (i. e., the SI system), the data and formulae employ that system of units.

The authors wish to express their gratitude to Prof. D. Moehlmann, Dr. sc. R. Waesch and Dr. sc. K. Sauer for helpful discussions and comments.

1. ORBITAL AND MECHANICAL PROPERTIES OF THE PLANETS, SATELLITES, PLANETARY RINGS, ASTEROIDS AND COMETS

1.1. Definitions and notations

[a]	- tropical year = 365.24139 d
[d]	- day(sidereal) = 24 hours =86400 s
[AU]	- astronomical unit, 1[AU] = 1.496E8 km
q	- distance of the perihelion from the Sun(hyperbolic orbits)
a,b	- semi-major and semi-minor axis of the orbit
$e=\sqrt{a^2-b^2}/a$	- eccentricity of the orbit (valid for elliptical orbits e < 1)
i, i_E, i_O	- inclination of the orbit to the ecliptic (latter both for satellites: to the planets equator i_E ; to the planets orbit i_O)
Node	- longitude of the ascending node of the orbit on the ecliptic, measured from the equinox
Peri	- argument of perihelion, measured from the node
P	- sidereal period = true period of the orbital revolution of a planet, satellite or asteroid arround its central body (outer of Mars)
$\mu = \frac{360°}{P[d]} * 1d$	- mean dayly motion for an assumed circular orbit of the body
m	- mass
$\bar{\rho}$	- mean density
$D, D_E, D_P,$	- diameter (latter both for planets: equatorial and polar diameter, respectively). In the case of satellites two or three given values for D correspond to the axis a rotanional and triaxial ellipsoid, respectively.
P_r	- sidereal rotation period
i_{OE}	- inclination of equator to the orbit

All angles are given in decimal degrees.

1.2. Orbital Elements of the Planets (data from [1])

The osculating element are avaveged versus the time intervall from 1978 Nov. 28 to 1990 Dec. 15 (for the terrestrial planets the range 1970 - 1990 is taken).
The mean values and their root mean square deviations are presented.

Planet	a $[10^6 \text{ km}]$	e	i	Node	P
Mercury	57.9	0.205630 +/- 3E-6	7.0044 +/- 3E-4	48.1 +/- 0.2	0.24085
Venus	108.2	0.006783 +/- 5E-6	3.3944 +/- 3E-4	76.5 +/- 0.1	0.61521
Earth	149.6	0.016718 +/- 5E-6	-	-	1.00004
Mars	227.9	0.093387 +/- 1E-5	1.8497 +/- 2E-4	49.4 +/- 0.1	1.88089
Jupiter	778.4 +/- 0.1	0.04803 +/- 2E-4	1.3053 +/- 3E-4	100.32 +/- 0.1	11.869
Saturn	1429.2 +/- 20	0.0537 +/- 3E-3	2.4867 +/- 2E-3	113.53 +/- 0.1	29.628
Uranus	2878.1 +/- 10	0.0481 +/- 3E-3	0.7725 +/- 3E-3	73.98 +/- 2E-2	4.665
Neptune	4508 +/- 20	0.008 +/- 3E-3	1.7714 +/- 2E-3	131.6 +/- 0.2	165.49
Pluto	5928 +/- 40	0.252 +/- 5E-3	17.139 +/- 1E-2	110.1 +/- 0.2	251.86
+)	5941.79 +/- 0.1	0.25234645	17.139029	109.513048	

+) - all data in this line taken from [7]

1.3. Mechanical Properties of the Planets (data taken from [1])

The negative value of P_R for Venus corresponds to a retrograde rotation.

Planet	m	$\bar{\rho}$	P_R	D_E	D_P	i_{OE}
	$[10^{24}$ kg]	$[10^{3}$ kgm^{-3}]	[d]	$[10^{3}$ km]	$[10^{3}$ km]	
Mercury	0.3302	5.43	58.65	4.878	4.878	2
Venus	4.869	5.24	-243.0	12.104	12.104	3
Earth	5.974	5.515	0.9972696	12.75628	12.71351	23.45
Mars	0.6419	3.93	1.025957	6.7944	6.7546	23.98
Jupiter	1898.8	1.33	0.4119 +/- 3E-3	142.796	133.800	3.067
Saturn	568.4	0.7	0.435 +/- 0.01	120.000	106.9	26.733
Uranus	86.98	1.27	(*)1.0 +/- 0.2	50.8	50.8	98
Neptune	102.8	1.71	(*)0.6 +/- 0.13	48.6	48.6	>29
Pluto	0.015	0.7	6.4	48.6	48.6	>50

(*) - data taken from [6]

1.4. Orbital Elements of the Satellites

The references are given below each value.

Planet	satellite	a $[10^{3}$ km]	e	i_E	i_0	P [d]
Earth	Moon	384.4 [1]	0.0549 [1]	18.3-28.6 [1]	5.1 [1]	27.321661 [1]
Mars	Phobos	9.38 [1]	0.015 [1]	1.1 [1]	[1]	0.3189 [1]
	Deimos	2346 [1]	0.00052 [1]	0.9 - 2.7 [1]	[1]	1.262 [1]
Jupiter	1979 J1	127.8 [1] 134 [3]	0.0 [1]	0.0 [1]		0.294 [1] 0.30 [3]
	Amalthea	181.3 [1,2]	0.0028 [1]	0.4 [1]		0.498 [1] 0.489 [2]
	1979 J2	222 [3] 221.7 [1]		1.25 [1]		0.67 [3] 0.675 [1]
	Io	422.0 [3] 421.6 [1,2]	0.0000 [1]	0.00 [1]		1.77 [3] 1.769 [1,2]

Planet	satellite	a [10^3 km]	e	i_E	i_O	P [d]
Jupiter	Europa	671 [3] 670.9 [1,2]	0.0003 [1]	0.02 [1]		3.55 [3] 3.551 [1,2]
	Ganymede	1070.0 [1,2,3]	0.0015 [1]	0.09 [1]		7.15 [3] 7.155 [1,2]
	Callisto	1180 [1,2,3]	0.0075 [1]	0.43 [1]		16.7 [3] 16.689 [1,2]
	Leda	11110 [3] 11094 [1]	0.148 [1]		27 [1]	240 [3] 239 [1]
	Himalia	11470 [1,3]	0.158 [1]		28 [1]	251 [3] 250.6 [1]
	Lysithea	11710 [1,3]	0.13 [1]		29 [1]	260 [1,3]
	Elara	11740 [1,3]	0.207 [1]		28 [1]	260 [3] 260.1 [1]
	Ananke	20700 [1,3]	0.17 [1]		147 [1]	617 [1,3]
	Carme	22350 [1,3]	0.21 [1]		163 [1]	692 [1,3]
	Pasiphae	23300 [1,3]	0.38 [1]		148 [1]	735 [1,3]
	Sinope	23700 [1,3]	0.28 [1]		153 [1]	758 [1,3]
Saturn	1980 S28	138.0 [1] 137.67 [8,9]		0 [1]		0.6 [1] 0.602 [8,9]
	1980 S27	140.0 [1] 139.35 [8,9]		0 [1]		0.61 [1] 0.613 [8,9]
	1980 S26	142.0 [1] 141.7 [8,9]		0 [1]		0.63 [1] 0.6285 [8,9]

Planet	satellite	a $[10^3$ km]	e	i_E	i_0	P [d]
Saturn	1980 S3 -	151.0 [1] 151.472 [8,9]	0.0 [1]	0 [1]		0.694 [1] 0.6947 [8,9]
	1980 S1	151.0 [1] 151.422 [8,9]		0 [1]		0.694 [1] 0.6944 [8,9]
	Mimas	≈ 186.0 [1] 185.54 [8,9]	0.0201 [1]	1.5 [1]		0.042 [1] 0.94242 [8,9]
	Mimas comp.	186.0 [8]				
	Encladus	238.0 [1] 238.04 [8,9]	0.0044 [1]	0 [1]		1.37 [1] 1.3702 [8,9]
	Tethys	295.0 [1] 294.67 [8,9]	0.0000 [1]	1.1 [1]		1.888 [1] 1.8878 [8,9]
	1980 S13	306.0 [1] 294.67 [8,9]				2.0 [1] 1.8878 [8,9]
	1980 S25	294.67 [8,9]				1.8878 [8,9]
	Dione	377.0 [1] 377.42 [8,9]	0.0022 [1]	0 [1]		2.737 [1] 2.7369 [8,9]
	1980 S6	377.0 [1] 378.06 [8,9]		0 [1]		2.736 [1] 2.7391 [8,9]
	Rhea	527.0 [1] 527.1 [8,9]	0.0010 [1]	0.4 [1]		4.518 [1] 4.5175 [8,9]
	Titan	1222.0 [1] 1221.86 [8,9]	0.0289 [1]	0.3 [1]		15.95 [1] 15.9454 [8,9]

Planet	satellite	a [10^3 km]	e	i_E	i_0	P [d]
Saturn	Hyperion	1481.0 [1,8,9]	0.1042 [1]	0.4 [1]		21.28 [1] 21.2766 [8,9]
	Iapetus	3560 [1] 3560.8 [8,9]	0.0283 [1]		18.4 [1]	79.33 [1] 79.3308 [8,9]
	Phoebe	12933 [1] 12954.0 [8,9]	0.1591 [1]		174.8 [1]	549.2 [1] 550.45 [8,9]
Uranus	1986 U7	49.7 +/- 0.05 [13] 49.3 [15]				0.33 [15]
	1986 U8	53.8 +/- 0.05 [13] 53.3 [15]				0.3717 [15]
	1986 U9	59.2 +/- 0.05 [13] 59.1 [15]				0.4325 [15]
	1986 U3	61.8 +/- 0.05 [13] 61.751 [16] 61.75 [14]				0.46375 [14] 0.46360 [16]
	1986 U6	62.7 +/- 0.05 [13,14] 62.745 [16]				0.47458 [14] 0.47469 [16]
	1986 U2	64.6 +/- 0.05 [13] 64.35 [14] 64.354 [16]				0.49292 [14] 0.49306 [16]
	1986 U1	66.1 +/- 0.05 [13] 66.9 [14,16]				0.51333 [14] 0.51314 [16]
	1986 U4	69.9 +/- 0.05 [13] 69.92 [14] 69.917 [16]				0.55875 [14] 0.55855 [16]

Planet	satellite	a [10^3 km]	e	i_E	i_O	P [d]
	1986 U5	75.3 +/- 0.05 [13] 75.1 [14] 75.14 [16]				0.62208 [14] 0.62225 [16]
	1985 U1	86.0 +/- 0.05 [13] 85.98 [14] 85.982 [16]				0.76183 [14] 0.76182 [16]
	Miranda	130.0 [1] 129.783 +/- 0.066 [13]	0.017 [1]	3.4 [1]		1.413 [1]
	Ariel	192.0 [1] 191.239 +/- 0.057 [13]	0.0028 [1]	0 [1]		2.520 [1]
	Umbriel	267.0 [1] 265.909 +/- 0.048 [13]	0.0035 [1]	0 [1]		4.144 [1]
	Titania	438.0 [1] 435.844 +/- 0.086 [13]	0.0024 [1]	0 [1]		8.706 [1]
	Oberon	586 [1] 582.596 +/- 0.071 [13]	0.0007 [1]	0 [1]		13.46 [1]
Neptune	Triton	354.0 [1] 324.12 [4]	0.00 [1,4]	160.0 [1,4]		5.877 [1]
	Neired	5510 [1] 5039 [4]	0.75 [1,4]		28.0 [1] 27.6 [4]	360.1 [1]
Pluto	Charon	20 [1,10]	0 [1,10]		120 +/- 5 [10]	6.387 [1] 6.3867 [10]

1.5. Mechanical Properties of Satellites

The references are given below each value.

Planet	satellite	D [km]	m [kg]	$\bar{\rho}$ [10^3 kgm^{-3}]	Discovery
Earth	Moon	3476 [1]	7.35E22 [1]	3.34 [1]	
Mars	Phobos	27 * 22 * 19 [1]			1877 Hall
		27 * 21.4 * 19.2 [18]	(9.9 +/- 1.1)E15 [18]	2 +/- 0.5 [18]	
	Deimos	15 * 12 * 11 [1,18]	(2.0 +/- 0.7)E15 [18]	1.9 +/- 0.7 [18]	1877 Hall
Jupiter	1979 J1	< 40 [1]			1979 Voyager 2
	Amalthea	240 [1]			1892 Bernard
	1979 J2	≈ 80 [1]			1980 Voyager 2
	Io	3650 [1]	8.92E22 [1]	3.5 [1]	1610 Galilei
		3640 +/- 20 [3,5,17]	(8.91 +/- 0.11)E22 [5]	3.52 [5]	
	Europa	3120 [1]	4.86E22 [1]	3.1 [1]	1610 Galilei
		3000 +/- 200 [5]	(4.87 +/- 0.11)E22 [5]	3.45 +/- 0.75 [5]	
		3130 +/- 30 [3,17]		3.0 [3]	

	Ganymede	5280 [1] 5270 +/- 50 [3,5]	1.489E23 [1] (1.49 +/- 0.01)E23	1.9 [1,3] 1.95 +/- 0.08 [5]	1610 Galilei
	Callisto	4870 [1] 4840 [3] 5000 +/- 200 [5]	1.063E23 [1] (1.065 +/- 0.032)E23 [5]	1.8 [1,3] 1.62 +/- 0.34 [5]	1610 Galilei
	Leda	≈ 10 [1]			1974 Kowal
	Himalia	170 [1]			1904 Perrine
	Lysithea	≈ 24 [1]			1938 Nicholson
	Elara	80 [1]			1905 Perrine
	Ananke	≈ 20 [1]			1951 Nicholson
	Carme	≈ 30 [1]			1938 Nicholson
	Pasiphae	≈ 36 [1]			1908 Helotte
	Sinope	≈ 28 [1]			1914 Nicholson
Saturn	1980 S28	≈100 [1] 40 * 20 [8,9]			1980 Voyager 1

Planet	satellite	D [km]	m [kg]	$\bar{\varrho}$ $[10^3\ kgm^{-3}]$	Discovery
Saturn	1980 S27	≈ 200 [1] 110 * 90 * 70 [8,9]			1980 Voyager 1
	1980 S26	250 [1] 110 * 90 * 70 [8,9]			1980 Voyager 1
	1980 S3	200 [1] 140 * 120 * 100 [8,9]			1980 Cruikshank
	1980 S1	140 * 70 [1] 220 * 200 * 160 [8,9]			1976 Fountain, Larson
	Mimas [8,9]	≈ 360 [1] 392 +/- 6 [9]	3.7E19 [1] (3.75 +/- 0.008)E19 [8]	≈ 1.5 [1] 1.44 +/- 0.18 1.19 +/- 0.05 [9]	1789 Herschel
	Encladus	≈ 600 [1] 500 +/- 20 [8,9]	8.4E19 [1] (8.4 +/- 3)E19 [9]	≈ 0.7 [1] 1.2 +/- 0.4 [9] 1.16 +/- 0.55 [8]	1789 Herschel

Tethys	1040 [1] 1060 +/- 20 [8,9]	6.3E20 [1] (7.55 +/- 0.9)E20 [9]	1.1 [1] 1.21 +/- 0.16 [8,9]	1684 Cassini
1980 S13	34 * 28 * 26 [8,9]			1980 Reitma, Smith, Larson
1980 S25	34 * 22 * 22 [8,9]			1980 Voyager 1
Dione	1000 [1] 1120 +/- 20 [8,9]	1.2E21 [1] (1.05 +/- 0.03)E21 [9]	2.2 [1] 1.43 +/- 0.06 [8,9]	1684 Cassini
1980 S6	≈ 80 [1] 36 * 32 * 30 [8,9]			1980 Laques, Lecachaux
Rhea	1600 [1] 1530 +/- 10 [8,9]	2.2E21 [1] (2.49 +/- 0.15)E21 [9]	1.0 [1] 1.33 +/-0.09 [8,9]	1672 Cassini
Titan	5800 [1] 5150 +/- 4 [8,9]	1.4E23 [1] 1.3457E23 [9]	1.3 [1] 1.88 +/- 0.01 [8,9]	1655 Huyghens
Hyperion	224 [1] 410 * 260 * 220 [8,9]			1848 Bond
Iapetus	1450 [1] 1460 +/- 20 [8,9]	2.8E21 [1] (1.88 +/- 0.12)E21 [9]	1.8 [1] 1.16 +/- 0.09 [8,9]	1671 Cassini

Planet	satellite	D [km]	m [kg]	$\bar{\rho}$ [10^3 kgm^{-3}]	Discovery
Saturn	Phoebe	≈ 240 [1] 220 +/- 20 [8,9]			1898 Pickering
Uranus	1986 U7	≈ 40 [13]			Voyager 2
	1986 U8	≈ 50 [13]			Voyager 2
	1986 U9	≈ 50 [13]			Voyager 2
	1986 U3	≈ 60 [13]			Voyager 2
	1986 U6	≈ 60 [13]			Voyager 2
	1986 U2	≈ 80 [13]			Voyager 2
	1986 U1	≈ 80 [13]			Voyager 2
	1986 U4	≈ 60 [13]			Voyager 2
	1986 U5	≈ 60 [13]			Voyager 2
	1985 U1	170 +/- 10 [13]			Voyager 2

	Miranda	≈ 300 [1,17] 484 +/- 10 [13]		1.26 +/- 0.39 [13]	1948 Kuiper
	Ariel	≈ 800 [1,17] 1160 +/- 10 [13]		1.65 +/- 0.30 [13]	1851 Lassell
	Umbriel	1190 +/- 20 [13] 500 [1,17]		1.44 +/- 0.28 [13]	1851 Lassell
	Titania	≈ 1000 [1,17] 1610 +/- 10 [13]		1.59 +/- 0.09 [13]	1787 Herschel
	Oberon	≈ 900 [1] 1550 +/- 20 [13]		1.50 +/- 0.10 [13]	1787 Herschel
Neptune	Triton	4400 [1] 3200 [17]	2E23 [1]	4.6 [1]	1846 Lassell
	Neired	≈ 300 [1,17]			1949 Kuiper
Pluto	Charon	<2000 >1200 [1]			1978 Christy

1.6. Planetary Ring Properties

All distances are measured from the planetary center.

1.6.1. Jovian Ring Data (data taken from [3])

	outer edge distance [10^3 km]	approximate width [km]	approximate thickness [km]
Jovian ring	128.3	6000	$<$ 30

1.6.2. Saturnian Ring Data

All data without reference statement are taken from [8].

Feature	inner edge distance [10^3 km]	outer edge distance [10^3 km]	mean optical depth
ring D	≈ 117.0		
ring C	72.8 [1] 74.4		0.08 - 0.15 [19]
ring B	91.6 [1] 91.9	117.0 [1] 117.4	1.2 - 1.9 [19]
Cassini-division	≈ 117.0 [1] ≈ 117.4	≈ 122.0 [1] ≈ 121.9	≈ 0.12 [19]
ring A	122.0 [1] 121.9	137.0 [1] 136.6	0.57 - 0.7 [19]
ring F	(*)140.4 [1] (*)143.3		
ring G	(*)170.0		
ring E	198.0 [1] 180.0	360.0 [1] 480.0	

(*) - distance corresponding to the centre of the ring

1.6.3. Uranian Ring Data

All data without reference statement are taken from [12].
The distances are with respect to the center of the rings.

	distance $[10^3 \text{ km}]$	e $[10^{-3}]$	i_E $[10^{-3}$ degrees$]$	width [km]	mean optical depth
1986 U 2R	37-39.5	≈ 0	≈ 0	≈ 2500	1E-3 - 1E-4
6 ring	42.02 [1] 41.98 [11] 41.85	1.0	63	1 - 3	0.2 - 0.3
5 ring	42.34 [1] 42.36 [11] 42.24	1.9	52	2 - 3	0.5 - 0.6
4 ring	42.66 [1] 42.663 [11] 42.58	1.1	32	2 - 3	0.3
α	44.83 [1] 44.844 [11] 44.73	0.8	14	7 - 12	0.3 - 0.4
β	45.80 [1] 45.799 [11] 45.67	0.4	5	7 - 12	0.2
η	47.31 [1] 47.323 [11] 47.18	≈ 0	≈ 2	0 - 2	0.1 - 0.4
γ	47.75 [1] 47.746 [11] 47.63	≈ 0	≈ 11	1 - 4	1.3 - 2.3
δ	48.42 [1] 48.423 [11] 48.31	≈ 0	≈ 4	3 - 9	0.3 - 0.4
1986 U 1R	50.04	0	0	1 - 2	0.1
	51.16	7.9	≈ 1	22 - 93	0.5 - 2.1

1.7. Orbital Elements of the Asteroids (data from [21])

Asteroid		a	e	i	Node	Peri	μ
	1950 DA	1.684	0.502	12.0	356.2	224.2	0.451
	1977 RA	2.104	0.413	5.24	287.0	42.2	0.322
	1977 VA	1.864	0.394	2.98	223.9	172.3	0.387
	1979 QB	2.332	0.440	3.35	341.9	11.8	0.276
	1979 VA	2.642	0.621	2.78	270.1	91.2	0.229
	1980 AA	1.892	0.443	4.18	298.3	168.1	0.378
	1980 PA	1.925	0.458	2.17	260.9	125.8	0.368
	1980 YS	1.815	0.321	2.27	58.5	49.1	0.403
	1981 CW	1.876	0.368	4.76	106.4	15.6	0.383
	1981 QA	2.152	0.448	8.41	171.6	154.3	0.312
	1982 DB	1.485	0.357	1.39	313.8	158.3	0.544
	1982 TA	2.297	0.771	12.20	9.3	119.4	0.282
	1982 XB	1.834	0.445	3.86	74.4	16.9	0.396
	1984 AB	1.577	0.075	14.84	94.8	115.4	0.497
	1984 KB	2.221	0.764	4.86	169.1	336.6	0.297
	1984 KD	2.203	0.544	13.64	81.7	203.6	0.301
433	Eros	1.458	0.223	10.828	303.771	178.540	0.560
887	Alinda	2.493	0.558	9.248	110.242	349.675	0.250
1065	Ammundsenia	2 361	0.298	8.373	329.984	352.998	0.272
1139	Atami	1.947	0.256	13.093	212.872	206.240	0.363
1221	Amor	1.921	0.434	11.892	170.887	26.178	0.370
1468	Zomba	2.196	0.271	9.953	308.544	22.512	0.303
1593	Fagnes	2.225	0.281	9.977	119.649	184.750	0.297
1620	Geographos	1.245	0.335	13.321	336.734	276.563	0.710
1627	Ivar	1.863	0.397	8.443	132.675	167.327	0.388
1640	Nemo	2.289	0.342	7.103	355.096	354.140	0.285
1685	Toro	1.367	0.436	9.374	273.790	126.821	0.617
1862	Apollo	1.471	0.560	6.351	35.403	285.463	0.552
1916	Boreas	2.273	0.450	12.845	340.300	335.305	0.288
1943	Anteros	1.430	0.256	8.703	245.785	338.104	0.576
2059	Baboquivari	2.652	0.526	10.993	200.485	191.189	0.228
2061	Anza	2.265	0.537	3.739	207.347	155.795	0.289
2062	Aten	0.966	0.183	18.935	108.050	147.856	1.037
2063	Bacchus	1.078	0.349	9.420	32.697	55.014	0.881
2064	Thomsen	2.179	0.329	5.695	301.776	2.108	0.306
2100	Ra-Shalom	0.832	0.436	15.760	170.294	355.894	1.299
2101	Adonis	1.875	0.764	1.362	350.669	41.579	0.384
2201	Oljato	2.173	0.712	2.516	76.435	95.730	0.308
2202	Pele	2.290	0.513	8.786	169.759	217.196	0.284
2212	Hephaistos	2.163	0.835	11.884	28.021	208.072	0.310
2253	Espinette	2.284	0.278	3.878	143.555	175.288	0.285
2340	Hathor	0.844	0.450	5.856	211.091	39.688	1.271
2368	Beltrovata	2.105	0.413	5.258	287.212	41.939	0.323
2608	Seneca	2.479	0.586	15.636	169.828	32.459	0.253
2744	Brigitta	2.303	0.332	6.763	315.325	43.253	0.282

1.8. Orbital Elements of the Comets (data taken from [20])

Comet (P/ - periodic comets)	perihelion time (*) year	month	day	q [AU]	e	i	Node	Peri	P [a]
N	-146	6	28.00000	0.430000	1.000000	71.0000	329.0000	261.0000	
N	240	11	10.00000	0.371000	1.000000	44.0000	213.0000	82.0000	
N	390	9	5.00000	0.920000	1.000000	36.0000	355.0000	23.0000	
N	400	2	25.00000	0.210000	1.000000	32.0000	37.0000	47.0000	
N	442	12	15.00000	1.530000	1.000000	106.0000	270.0000	178.0000	
N	539	11	6.00000	0.160000	1.000000	19.0000	32.0000	246.0000	
N	565	7	15.00000	0.832000	1.000000	121.0000	179.0000	79.0000	
N	568	8	27.70000	0.870000	1.000000	4.0000	301.0000	35.0000	
N	574	3	25.00000	0.730000	1.000000	54.0000	154.0000	342.0000	
N	770	6	5.80000	0.580000	1.000000	117.0000	110.0000	88.0000	
N	868	3	4.00000	0.420000	1.000000	65.0000	320.0000	277.0000	
N	905	4	26.00000	0.200000	1.000000	140.0000	69.0000	50.0000	
N	961	12	28.00000	0.630000	1.000000	119.0000	1.0000	85.0000	
N	1014	4	6.00000	0.560000	1.000000	117.0000	173.0000	84.0000	
N	1018	8	27.00000	0.620000	1.000000	35.0000	184.0000	197.0000	
N	1080	9	10.94000	0.680900	1.000000	6.8900	322.2000	73.6300	
N	1092	2	22.00000	0.770000	1.000000	124.0000	113.0000	62.0000	
N	1097	9	22.30000	0.302000	1.000000	41.0000	351.0000	298.0000	
N	1110	5	18.00000	0.830000	1.000000	136.5000	320.0000	358.0000	
N	1132	8	30.70000	0.735500	1.000000	106.3000	212.5000	114.2000	
N	1147	1	28.00000	0.120000	1.000000	110.0000	281.0000	300.0000	
N	1230	12	28.80000	0.860000	1.000000	16.0000	303.0000	181.0000	
N	1240	1	21.56000	0.667700	1.000000	75.3000	134.4000	331.4000	
N	1245	4	1.00000	0.500000	1.000000	20.0000	179.0000	87.0000	
N	1264	7	20.29000	0.824900	1.000000	16.4000	150.3500	159.7000	
N	1293	10	28.00000	0.780000	1.000000	30.0000	88.0000	313.0000	
N	1299	3	31.81000	0.317930	1.000000	111.0100	116.2400	103.8800	
N	1305	1	19.00000	0.650000	1.000000	99.0000	105.0000	56.0000	
N	1337	6	14.85000	0.749000	1.000000	143.6000	96.9000	79.6000	
N	1340	5	13.62000	1.136200	1.000000	176.4900	186.9500	32.4600	
N	1345	8	23.00000	0.890000	1.000000	23.0000	149.0000	210.0000	
N	1351	11	19.00000	1.010000	1.000000	7.0000	263.0000	165.0000	
N	1362	2	25.00000	0.300000	1.000000	155.0000	290.0000	348.0000	
N	1368	5	5.00000	0.480000	1.000000	131.0000	216.0000	90.0000	

Comet	perihelion time (*)			q	e	i	Node	Peri	P
(P/ - periodic comets)	year	month	day	[AU]					[a]
N	1376	7	31.70000	0.550000	1.000000	76.0000	300.0000	81.0000	
N	1385	10	24.00000	0.790000	1.000000	103.0000	288.0000	182.0000	
N	1402	3	21.00000	0.380000	1.000000	55.0000	125.0000	91.0000	
N	1433	11	8.27000	0.492770	1.000000	103.9810	103.5550	189.3860	
N	1439	5	9.00000	0.120000	1.000000	81.0000	191.0000	140.0000	
N	1449	12	9.86800	0.327420	1.000000	155.6790	268.1250	356.7070	
N	1457	1	18.47900	0.703360	1.000000	13.2520	256.8030	194.6250	
N	1457	8	8.10500	0.769910	1.000000	7.3420	186.8420	183.2920	
N	1458	11	7.00000	0.490000	1.000000	69.0000	116.0000	258.0000	
N	1462	8	4.00000	0.250000	1.000000	130.0000	315.0000	137.0000	
N	1468	10	7.30000	0.850000	1.000000	138.0000	106.0000	91.0000	
N	1472	3	1.43900	0.485890	1.000000	170.8630	292.2100	245.7640	
N	1491	1	8.90000	0.761000	1.000000	73.4000	279.5000	164.9000	
N	1491	1	22.70000	0.850000	1.000000	9.0000	165.0000	316.0000	
N	1499	9	9.00000	0.950000	1.000000	16.0000	328.0000	42.0000	
N	1500	4	30.00000	1.110000	1.000000	99.0000	304.0000	352.0000	
N	1506	8	28.50000	0.840000	1.000000	115.0000	155.0000	200.0000	
N	1532	10	18.83200	0.519220	1.000000	32.5900	93.1200	24.5200	
N	1533	6	15.41700	0.254800	1.000000	149.5960	124.7170	21.3880	
N	1537	12	30.32000	0.147700	1.000000	42.4500	355.5000	287.7000	
N	1539	5	12.60000	0.961000	1.000000	18.6000	54.4000	174.5000	
N	1556	4	22.68500	0.490820	1.000000	32.3770	180.7360	100.8680	
N	1557	9	22.00000	0.500000	1.000000	42.0000	211.0000	94.0000	
N	1558	9	14.04000	0.280500	1.000000	110.9300	340.5100	119.6000	
N	1577	10	27.44800	0.177500	1.000000	104.8780	30.5380	255.6690	
N	1577	11	8.50000	0.877000	1.000000	9.7000	35.4000	27.1000	
N	1580	11	28.99300	0.602370	1.000000	64.6060	24.2510	89.3630	
N	1582	5	6.91300	0.168720	1.000000	118.5430	232.3380	331.9010	
N	1585	10	8.52620	1.094846	1.000000	6.1302	42.4901	331.7326	
N	1590	2	8.52710	0.567726	1.000000	150.4570	170.6420	307.6700	
N	1593	7	19.06900	0.089113	1.000000	87.9200	169.2200	12.0700	
N	1596	7	25.71430	0.567156	1.000000	128.0740	335.2660	59.4170	
N	1618	8	17.62700	0.512979	1.000000	21.4900	298.1300	24.8200	
N	1618	11	8.85070	0.389544	1.000000	37.1956	80.3114	287.4248	
N	1652	11	13.15300	0.847500	1.000000	79.4600	92.3000	300.1800	

N	1661	1	27.38100	0.442722	1.000000	33.0150	85.8730	33.4380	
N	1664	12	4.98340	1.025531	1.000000	158.6968	85.3373	310.6569	
N	1665	4	24.71900	0.106490	1.000000	103.9000	232.0000	156.0900	
N	1668	2	28.08000	0.066604	1.000000	144.3750	2.5150	109.8120	
N	1672	3	1.94670	0.695458	1.000000	82.9677	301.9913	109.5297	
N	1677	5	6.52600	0.280590	1.000000	100.9310	240.6190	99.1680	
N	1678	8	27.08500	1.238020	1.000000	3.0400	165.3400	166.2100	
N	1680	12	18.48760	0.006222	0.999986	60.6771	275.9317	350.6202	
N	1683	7	13.59070	0.559584	1.000000	96.7443	177.1369	87.8099	
N	1684	6	8.76300	0.958270	1.000000	65.4230	271.8960	330.3070	
N	1686	9	16.32490	0.336020	1.000000	34.9630	357.7450	81.9130	
N	1689	11	30.65890	0.064430	1.000000	63.2020	283.0520	78.1410	
N	1695	10	23.76800	0.042297	1.000000	93.5870	285.3200	59.1240	
N	1698	10	17.51490	0.728650	1.000000	169.0900	69.5500	151.3500	
N	1699	1	13.89980	0.748620	1.000000	109.4190	325.1890	109.5280	
N	1701	10	17.91000	0.592630	1.000000	138.3700	302.1200	164.9600	
N	1702	3	14.10700	0.646830	1.000000	4.3810	192.5700	309.6630	
N	1706	1	30.70600	0.426865	1.000000	55.2640	16.5840	59.4320	
N	1707	12	12.48830	0.859040	1.000000	88.6430	56.2180	27.1560	
N	1718	1	15.40570	1.025435	1.000000	148.8430	131.1973	6.3002	
N	1723	9	28.12790	0.998790	1.000000	130.0220	17.4020	331.3760	
N	1729	6	16.64770	4.050540	1.000000	77.0900	313.6937	10.4077	
N	1737	1	30.84700	0.222823	1.000000	18.3300	229.4100	99.4800	
N	1737	6	3.53000	0.838100	1.000000	63.8400	135.9900	129.9600	
N	1739	6	17.91600	0.673578	1.000000	124.2650	210.3480	104.7570	
N	1742	2	8.69640	0.765769	1.000000	112.9541	188.5031	328.0445	
N	1743	1	11.34740	0.838181	1.000000	2.2780	69.7620	26.0920	
N	1743	9	21.14690	0.522993	1.000000	134.4080	8.9330	119.0450	
N	1744	3	1.83980	0.222209	1.000000	47.1378	48.6030	151.4783	
N	1747	3	3.91550	2.198590	1.000000	100.8630	150.1500	230.2880	
N	1748	4	29.30930	0.840671	1.000000	94.5380	235.6830	17.8340	
N	1748	6	19.38750	0.625357	1.000000	67.0780	35.9480	245.6640	
N	1750	2	23.00000	0.200000	1.000000	132.0000	303.0000	285.0000	
N	1757	10	21.82620	0.337376	1.000000	12.8329	216.9591	268.7099	
N	1758	6	11.63700	0.215352	1.000000	68.3000	233.5100	36.7800	
MESSIER	1759	11	27.50180	0.801390	1.000000	79.0900	142.3090	273.9240	
GREAT COMET	1759	12	17.34110	0.965760	1.000000	175.1260	82.7780	301.6510	
KLINKENBERG	1762	5	28.83450	1.009048	1.000000	85.6610	351.1700	115.4800	
MESSIER	1763	11	2.37380	0.498185	1.000000	72.4910	359.0480	88.5700	
MESSIER	1764	2	13.07100	0.555216	1.000000	127.0930	122.6850	104.8520	
MESSIER	1766	2	17.86160	0.505324	1.000000	139.1540	246.7190	100.8890	
P/HELFENZRIEDER	1766	4	27.90880	0.406034	0.847632	7.8635	75.6114	178.6515	4.35

Comet	perihelion time (*)			q	e	i	Node	Peri	P
(P/ - periodic comets)	year	month	day	[AU]					[a]
MESSIER	1769	10	8.12040	0.122755	0.999249	40.7403	177.5941	329.1231	
GREAT COMET	1770	11	22.73520	0.528242	1.000000	148.5580	111.2360	260.3640	
P/LEXELL(MESSIER)	1770	8	14.04090	0.674449	0.786119	1.5570	133.9260	224.8590	5.60
MESSIER	1771	4	19.63490	0.901835	1.000000	11.2838	30.3167	76.2057	
MESSIER	1773	9	6.10680	1.126886	1.000000	61.2240	123.5400	314.1120	
MONTAIGNE	1774	8	15.94880	1.428580	1.000000	82.9840	183.2760	136.5340	
BODE	1779	1	4.58070	0.713178	1.000000	32.4351	27.4438	62.1678	
MESSIER	1780	10	1.25240	0.099256	1.000000	126.1810	126.5290	237.8210	
MONTAIGNE-OLBERS	1780	11	29.34800	0.515277	1.000000	107.9200	144.3800	255.1600	
MECHAIN	1781	7	7.68890	0.775861	1.000000	81.7240	85.3590	156.2020	
MECHAIN	1781	11	30.02320	0.960995	1.000000	152.8010	79.7710	61.3780	
P/PIGOTT	1783	11	20.43040	1.459289	0.552456	45.1248	57.9867	354.6428	5.89
GREAT COMET	1784	1	21.69960	0.707858	1.000000	128.8560	59.1530	336.1080	
MESSIER-MECHAIN	1785	1	27.82550	1.143398	1.000000	70.2370	266.5140	205.6380	
MECHAIN	1785	4	8.91400	0.427240	0.996461	92.6367	66.9888	127.1968	
HERSCHEL	1786	7	8.37970	0.410712	1.000000	50.8981	196.7000	325.0548	
MECHAIN	1787	5	11.32500	0.348911	1.000000	131.7270	109.1480	99.1500	
MESSIER	1788	11	10.80930	1.063012	1.000000	167.5190	159.2200	57.8360	
HERSCHEL	1790	1	15.71200	0.758097	1.000000	148.0750	178.4270	115.9510	
HERSCHEL	1790	5	21.74100	0.797961	1.000000	116.1420	35.4170	119.4750	
HERSCHEL	1792	1	14.03490	1.292577	1.000000	140.2170	192.9000	154.3500	
GREGORY	1792	12	27.69800	0.965812	1.000000	131.0010	285.4670	147.1750	
MESSIER	1793	11	5.34100	0.403441	1.000000	119.6440	110.6810	239.8030	
PERNY	1793	11	19.99830	1.504546	1.000000	51.9298	4.4773	69.3347	
OLBERS	1796	4	3.27370	1.579636	1.000000	115.1657	19.1978	184.2940	
BOUVARD-HERSCHEL-LEE	1797	7	9.61420	0.525450	1.000000	129.4210	331.3970	279.6860	
MESSIER	1798	4	5.00830	0.485030	1.000000	43.7880	124.2280	342.9950	
BOUVARD	1798	1	1.41380	0.774794	1.000000	137.7480	251.5880	215.8870	
MECHAIN	1799	9	7.69090	0.839863	1.000000	129.0630	101.6230	95.8200	
MECHAIN	1799	12	26.39640	0.625802	1.000000	102.9900	328.9120	136.4740	
PONS	1801	8	9.05650	0.256410	1.000000	159.2630	44.6010	219.8260	
PONS	1802	9	10.35540	1.094249	1.000000	57.0060	312.3128	21.8530	
PONS	1804	2	14.08810	1.071168	1.000000	56.4590	178.8360	331.9460	
PONS	1806	12	29.41180	1.081930	1.000000	144.9480	324.2930	225.2190	
GREAT COMET	1807	9	19.23890	0.646124	0.995488	63.1758	268.7819	4.1043	

PONS	1808	5	13.45300	0.389860	1.000000	134.2970	324.9460	253.7480	
PONS	1808	7	12.66770	0.607953	1.000000	140.6990	26.1740	131.5550	
PONS	1810	10	6.73790	0.969623	1.000000	62.9407	310.8026	114.9226	
GREAT COMET	1811	9	12.75620	1.035412	0.995125	106.9397	142.3501	65.4060	
PONS	1811	11	11.54270	1.581701	0.980916	31.2565	94.9426	314.4898	755.00
PONS	1813	3	5.02200	0.699314	1.000000	158.8511	242.4627	170.5796	
PONS	1813	5	20.08720	1.214703	1.000000	99.0530	44.5760	205.1570	
PONS	1816	3	1.84600	0.048503	1.000000	43.1060	325.1270	304.3310	
PONS	1818	2	26.46900	1.197821	1.000000	89.7504	72.2965	112.3556	
PONS	1818	12	5.43470	0.855096	1.000000	116.9070	91.8350	348.1000	
GREAT COMET(TRALLES)	1819	6	28.21810	0.341514	1.000000	80.7504	275.5353	13.4223	
P/BLANPAIN	1819	11	20.84740	0.892318	0.698752	9.1074	79.1541	350.2203	5.10
NICOLLET-PONS	1821	3	22.03660	0.091823	1.000000	106.4580	50.4840	169.2060	
PONS	1822	7	16.33370	0.847130	1.000000	143.6930	99.5340	237.7650	
PONS	1822	10	24.26570	1.145099	0.996316	127.3440	94.5391	181.0974	
GAMBART	1822	5	6.07930	0.504430	1.000000	126.4030	179.2360	344.6910	
GREAT COMET	1823	12	9.93400	0.226742	1.000000	103.8152	304.8081	28.4919	
RUEMKER	1824	7	12.01880	0.591671	1.000000	125.4355	236.1008	334.0870	
SCHEITHAUER	1824	9	29.56640	1.049835	1.000000	54.5967	281.0385	85.2368	
PONS	1825	8	19.21110	0.883471	1.000000	89.6810	194.6809	177.2987	
PONS	1825	12	11.18500	1.240846	0.995431	146.4401	217.4377	256.9254	
GAMBART	1825	5	31.02260	0.889011	0.996395	123.3356	21.8827	106.1874	
PONS	1826	4	22.45810	2.007734	1.000000	39.9962	199.3572	279.3883	
PONS	1826	10	9.47610	0.852878	0.997492	25.9455	45.7143	13.7985	
PONS	1826	11	18.90740	0.026904	1.000000	90.6265	236.8520	279.5892	
PONS	1827	2	5.41100	0.506166	1.000000	102.3962	186.2942	151.0353	
PONS	1827	9	12.16700	0.137758	1.000000	125.8847	151.4003	258.7007	
P/PONS-GAMBART	1827	6	7.63760	0.806508	0.945838	136.4548	319.3346	19.1950	57.50
GREAT COMET(DABADIE)	1830	4	9.79510	0.921424	1.000000	21.2625	208.0565	5.8060	
GREAT COMET(HERAPATH)	1830	12	28.16040	0.125887	1.000000	135.2567	339.5424	26.8909	
GAMBART	1832	9	26.07350	1.183005	1.000000	136.6715	74.1216	204.6248	
DUNLOP	1833	9	10.69120	0.458122	1.000000	7.3432	324.8636	259.6049	
GAMBART	1834	4	3.29260	0.513106	1.000000	5.9798	228.2891	50.0380	
BOGUSLAWSKI	1835	3	27.70520	2.040153	1.000000	170.8819	60.0327	210.5257	
GALLE	1840	1	4.97520	0.618434	1.000000	53.0755	121.4842	72.2606	
GALLE	1840	3	13.62120	1.220451	0.993250	120.7836	238.3708	156.6266	
GALLE	1840	4	2.93780	0.748504	1.000000	79.8576	187.5727	138.0454	
BREMIKER	1840	11	14.06520	1.479951	0.965941	57.9059	250.5120	133.5202	286.00
LAUGIER	1842	12	16.45530	0.504685	1.000000	106.4267	209.3114	240.4976	
GREAT MARCH COMET	1843	2	27.91100	0.005527	0.999914	144.3484	2.8274	82.6374	513.00
MAUVAIS	1843	5	6.55590	1.616305	1.000000	52.7304	158.7439	124.2504	
GREAT COMET(WILMOT)	1844	12	14.19140	0.250537	0.999302	45.5689	119.8977	177.4980	

Comet (P/ - periodic comets)	perihelion time (*) year	month	day	q [AU]	e	i	Node	Peri	P [a]
MAUVAIS	1844	10	17.84600	0.855401	0.999757	131.4041	33.1462	211.2694	
GREAT JUNE COMET(COLLA)	1845	6	6.18320	0.401077	1.000000	131.0922	339.2759	75.8000	
D´ARREST	1845	1	8.65790	0.905204	1.000000	46.8543	338.2132	114.5845	
DE VICO	1845	4	21.54040	1.254549	1.000000	56.3973	348.5821	205.4524	
P/DE VICO	1846	3	6.04460	0.663802	0.963099	85.1128	79.0127	12.8995	76.30
DE VICO	1846	1	22.59390	1.480703	0.992403	47.4287	112.5813	337.9820	
DE VICO	1846	10	30.27720	0.830669	1.000000	49.7132	6.1395	93.9746	
DE VICO-HIND	1846	5	28.39580	1.375992	1.000000	122.3835	162.7648	78.7501	
P/PETERS	1846	6	1.66060	1.526573	0.719689	30.5378	261.9405	339.6181	12.70
BRORSEN	1846	6	5.97100	0.633760	0.990414	150.6811	263.3020	99.7386	538.00
MAUVAIS	1847	8	9.83860	1.766058	0.998589	96.5754	339.7061	91.5265	
HIND	1847	3	30.78440	0.042593	0.999910	48.6579	23.1282	254.3524	
COLLA	1847	6	5.22980	2.115749	1.000723	100.4250	175.3864	32.3588	
SCHWEIZER	1847	8	9.84510	1.484824	1.000000	147.3544	78.1749	55.4597	
MITCHELL	1847	11	14.89990	0.329025	1.000172	108.1388	192.2686	276.6110	
PETERSEN	1848	9	8.54530	0.319947	1.000000	95.5989	212.9517	260.9518	
SCHWEIZER	1849	6	8.71220	0.894360	0.998427	66.9535	31.9382	236.5825	
PETERSEN	1849	1	19.85340	0.959782	1.000000	85.0449	216.6343	208.0110	
GOUJON	1849	5	26.99230	1.159406	1.000940	67.1583	203.9603	33.1655	
PETERSEN	1850	7	24.02860	1.081429	0.998599	68.1859	94.2873	180.5426	
BOND	1850	10	19.83740	0.565586	1.000000	40.0676	207.4085	243.2084	
BRORSEN	1851	8	26.74580	0.984753	1.000000	38.2077	225.0673	87.2684	
BRORSEN	1851	10	1.29620	0.142053	1.000000	73.9821	45.7318	294.4418	
P/BIELA(SECCHI)	1852	9	23.54320	0.860594	0.755828	12.5507	247.2805	223.2176	6.62
CHACORNAC	1852	4	20.08730	0.905406	1.000000	131.1191	318.5760	37.2112	
KLINKERFUES	1853	9	2.20690	0.306834	1.000250	61.5063	141.8698	170.4372	
SCHWEIZER	1853	5	10.32630	0.908693	0.989286	122.1911	42.3203	199.2278	781.00
SECCHI	1853	2	24.52130	1.092195	1.000000	159.7550	70.9417	275.8868	
BRUHNS	1853	10	17.12440	0.172863	1.000664	119.0008	221.4202	277.8461	
GREAT COMET	1854	3	24.51320	0.277064	1.000000	97.4816	316.7980	101.6296	
KLINKERFUES	1854	6	22.49700	0.648092	1.000000	108.6967	349.0001	74.5696	
KLINKERFUES	1854	10	28.00850	0.798762	0.993246	40.9144	325.8140	129.9036	
VAN ARSDALE	1854	1	4.43510	2.044650	1.000000	113.8796	228.3823	170.9259	
WINNECKE-DIEN	1854	12	16.30920	1.359463	0.991309	14.1548	239.5343	287.0966	
SCHWEIZER	1855	2	5.54740	2.193526	0.965185	128.5827	191.0492	323.0952	500.00

BRUHNS	1855	11	25.88720	1.230980	1.000000	169.8206	52.9322	325.5839	
DONATI	1855	5	30.65400	0.567564	0.985780	156.8710	261.5473	22.5049	252.00
KLINKERFUES	1857	7	18.47430	0.367535	1.000000	121.0276	24.9939	134.0619	
KLINKERFUES	1857	10	1.37990	0.562900	0.996913	123.9553	16.2549	124.8415	
D'ARREST	1857	3	21.86910	0.772493	1.000000	87.9430	314.4539	121.5706	
PETERS	1857	8	24.49640	0.746843	0.980414	32.7623	202.1281	180.9495	235.00
DONATI-VAN ARSDALE	1857	11	19.56410	1.009088	1.000000	142.1611	140.6041	95.0961	
BRUHNS	1858	6	5.79560	0.544261	1.000000	99.9655	326.2523	98.8618	
DONATI	1858	9	30.46450	0.578469	0.996295	116.9577	166.6054	129.1134	
TUTTLE	1858	10	13.31020	1.427002	1.000000	158.7114	161.0423	155.5369	
TEMPEL	1859	5	29.72610	0.201032	1.000000	95.4829	358.6160	282.0045	
GREAT COMET	1860	6	16.56100	0.292885	1.000000	79.3235	85.9298	76.8774	
TEMPEL	1860	9	22.81190	0.682650	1.000000	32.2060	46.0840	311.9670	
LIAIS	1860	2	17.12490	1.198875	1.000000	79.6762	325.3187	209.7632	
RUMKER	1860	3	6.06550	1.306664	1.000000	48.2297	10.1291	41.2149	
GREAT COMET(TEBBUTT)	1861	6	12.00680	0.822384	0.985070	85.4407	280.2109	330.0904	409.00
TUTTLE	1861	12	7.67440	0.839027	1.000000	138.0076	146.3353	331.5927	
THATCHER	1861	6	3.88990	0.920700	0.983465	79.7681	31.1696	213.4457	415.00
SCHMIDT	1862	6	22.52960	0.981328	1.000000	172.1033	327.7599	27.1880	
P/SWIFT-TUTTLE	1862	8	23.40970	0.962638	0.960427	113.5596	138.6849	152.7660	120.00
RESPIGHI	1862	12	28.67410	0.803238	1.000000	137.5347	356.9964	230.5760	
KLINKERFUES	1863	4	5.39610	1.068038	0.999159	112.6223	252.4702	3.9738	
BRUHNS	1863	2	3.99020	0.794764	1.000049	85.3589	118.1406	74.4602	
TEMPEL	1863	11	9.97180	0.706413	0.998879	78.0833	98.6968	357.2064	
RESPIGHI	1863	4	21.36460	0.628781	0.999079	85.4976	251.3834	55.5989	
RESPIGHI	1863	12	28.26270	0.771492	1.000000	64.4863	305.9278	115.6746	
BAEKER	1863	12	29.66770	1.313114	1.000665	83.3174	106.2368	78.1105	
BRUHNS	1864	12	28.21880	1.114634	1.000000	162.8868	342.0850	178.5042	
DONATI	1864	7	28.31180	0.626106	1.000000	134.9887	176.1824	346.0951	
DONATI-TOUSSAINT	1864	10	11.89210	0.931212	0.999358	109.7072	32.9662	232.4550	
TEMPEL	1864	8	16.07670	0.909290	0.996351	178.1283	96.7806	151.3851	
BAEKER	1864	12	22.95110	0.770731	1.000000	48.8680	204.4112	118.4565	
GREAT SOUTHERN COMET	1865	1	14.82530	0.025844	1.000000	92.4957	254.1269	111.7239	
BAEKER-WINNECKE	1867	11	7.46080	0.330359	1.000000	96.5714	66.1425	148.6346	
WINNECKE	1868	6	26.97630	0.578577	1.000000	131.5547	53.4089	126.6357	
TEMPEL	1869	10	10.34960	1.230731	1.000000	111.6805	312.6309	188.1987	
WINNECKE-TEMPEL	1870	7	14.58220	1.008692	1.000000	121.7890	142.8668	198.2238	
WINNECKE	1870	12	20.37610	0.389262	1.000000	147.2716	95.8781	90.6155	
COGGIA	1870	9	2.66760	1.816602	1.000000	99.3518	14.0602	354.9367	
WINNECKE	1871	6	11.09990	0.654300	0.997814	87.6017	280.4134	222.5131	
TEMPEL	1871	7	27.53440	1.083363	1.000000	101.9833	213.0053	96.3261	
TEMPEL	1871	12	20.87220	0.691268	0.995714	98.3051	148.2301	242.8905	

Comet (P/ - periodic comets)	perihelion time (*) year	month	day	q [AU]	e	i	Node	Peri	P [a]
BORRELLY	1873	9	11.28370	0.794061	0.996482	95.9698	231.6641	193.7784	
HENRY	1873	10	2.26720	0.384913	0.999730	121.4691	177.7976	233.7548	
WINNECKE	1874	3	10.43460	0.044568	1.000000	58.8880	31.3579	269.5053	
WINNECKE	1874	3	14.43550	0.885773	1.000000	148.4110	275.1605	331.7289	
COGGIA	1874	7	9.35830	0.675782	0.998820	66.3476	119.7988	152.3745	
COGGIA	1874	7	18.20160	1.687985	0.962822	34.1315	216.9224	149.5928	306.00
BORRELLY	1874	8	27.34360	0.982649	0.998831	41.8280	252.5741	92.6219	
BORRELLY	1874	10	19.44280	0.508226	1.000000	99.2185	283.0189	16.2745	
WINNECKE	1877	4	18.15620	0.949980	0.998700	121.1496	317.6379	63.1250	
TEMPEL	1877	6	27.57030	1.070451	1.000000	115.7325	185.3002	103.2450	
COGGIA	1877	9	11.71820	1.575904	1.000000	102.2289	252.0134	143.2115	
BORRELLY	1877	1	19.67860	0.807441	1.000000	152.9011	188.2653	347.1645	
SWIFT-BORRELLY-BLOCK	1877	4	27.30580	1.009053	0.997923	77.1852	347.0997	116.7730	
SWIFT -	1878	7	21.19070	1.391967	1.000000	78.1781	103.2672	177.5835	
P/BRORSEN(TEMPEL)	1879	3	31.03450	0.589846	0.809820	29.3845	102.2793	14.9348	5.46
SWIFT	1879	4	27.92250	0.896547	1.000000	107.0400	46.7545	3.7475	
HARTWIG	1879	8	29.77870	0.991418	1.000000	107.7578	33.4189	84.2591	
PALISA	1879	10	5.12500	0.989608	1.000000	77.1305	88.1718	115.4493	
GREAT SOUTHERN COMET(GOULD)	1880	1	28.11820	0.005494	1.000000	144.6603	7.0769	86.2462	
HARTWIG	1880	9	7.43510	0.354634	1.000000	141.9079	46.3014	323.1205	
SCHAEBERLE	1880	7	2.20380	1.814412	1.000813	123.0570	258.2304	145.1753	
PECHULE	1880	11	9.92000	0.659736	1.000000	60.7023	250.3474	11.6949	
GREAT COMET(TEBBUTT)	1881	6	16.94070	0.734558	0.995946	63.4244	271.9290	354.2422	
SWIFT	1881	5	20.94090	0.591090	1.000000	77.9380	127.4097	173.8008	
SWIFT	1881	11	20.24180	1.924973	0.990169	144.8080	182.3421	117.9503	
SCHAEBERLE	1881	8	22.80600	0.633542	1.000000	140.2296	98.0175	122.1356	
BARNARD	1881	9	14.86530	0.449200	1.000000	112.8148	275.1237	6.2930	
GREAT SEPTEMBER COMET(CRULS)	1882	9	17.72410	0.007751	0.999907	142.0046	346.9592	69.5867	759.00
BARNARD	1882	11	13.47540	0.955567	1.000000	96.1482	250.0724	254.2947	
WELLS	1882	6	11.02960	0.060763	0.999994	73.8033	205.8916	208.9889	
BROOKS-SWIFT	1883	2	19.45330	0.760062	1.000000	78.0632	279.0736	110.9011	
ROSS	1883	12	25.79260	0.309607	1.000000	114.9769	265.2970	138.6040	
P/BARNARD 1	1884	8	16.96800	1.279482	0.583252	5.4637	6.0750	301.0374	5.38
BARNARD	1885	8	6.18180	2.506732	1.000000	80.6249	93.2021	178.5098	
BROOKS	1885	8	10.65730	0.749149	0.982264	59.1026	205.6669	42.8540	275.00

BROOKS	1885	11	26.00500	1.079545	1.000000	42.4424	263.1057	35.6001	
BARNARD	1886	5	3.78560	0.479259	1.000238	84.4397	69.2121	119.6167	
BARNARD	1886	11	28.90560	1.480714	1.000000	85.5874	259.0972	31.9180	
BROOKS	1886	5	4.95100	0.841990	1.000000	100.2150	288.6680	38.5650	
BROOKS	1886	6	7.88540	0.269803	0.996783	87.6677	193.5059	201.2893	768.00
FABRY	1886	4	6.45420	0.642363	1.000272	82.6238	37.2652	126.5833	
P/BROOKS 1	1886	6	7.21220	1.325273	0.571388	12.6677	54.4656	176.8499	5.44
GREAT SOUTHERN COMET(THOME)	1887	1	11.93400	0.004834	1.000000	144.3770	3.8850	83.5110	
BARNARD	1887	3	28.93590	1.006455	1.000000	139.7814	136.3386	36.5476	
BARNARD	1887	6	17.16300	1.393813	0.996093	17.5500	246.1270	15.1081	
BROOKS	1887	3	17.89070	1.630145	0.983694	104.2720	280.8144	159.4296	
BARNARD	1888	9	13.28240	1.527853	0.991488	56.3478	138.3936	290.8040	
BROOKS	1888	7	31.63640	0.902226	0.999908	74.1923	102.3608	59.2103	
SAWERTHAL	1888	3	17.50210	0.698772	0.995874	42.2503	246.2570	359.9141	
BARNARD	1889	1	31.66820	1.814916	1.001230	166.3777	358.2749	340.4625	
BARNARD	1889	6	11.31390	2.255596	0.999818	163.8470	311.5380	236.0778	
P/BARNARD 2	1889	6	21.24880	1.104894	0.960033	31.2460	271.9250	60.2446	145.00
DAVIDSON	1889	7	19.78400	1.039721	0.997611	65.9891	287.0103	345.8685	
COGGIA	1890	7	9.03730	0.764087	1.000000	63.3447	15.1373	85.6583	
BORRELLY	1890	1	26.97640	0.269796	1.000000	56.7474	9.3208	199.8531	
BROOKS	1890	6	2.03750	1.907582	1.000266	120.5636	321.1815	68.9315	
ZONA	1890	8	7.38270	2.046690	0.995872	154.3072	86.2221	331.2691	
DENNING	1890	9	25.00780	1.260223	0.999187	98.9391	100.9594	163.0460	
P/SPITALER	1890	10	27.09810	1.817661	0.471181	12.8420	45.8949	13.3879	6.37
BARNARD	1891	11	14.04270	0.971089	1.000000	77.9927	218.8281	269.5717	
BARNARD-DENNING	1891	4	28.02100	0.397896	1.000000	120.5177	194.7563	178.7527	
SWIFT	1892	4	7.15170	1.026832	0.998731	38.7028	241.7237	24.5070	
BROOKS	1892	12	28.58640	0.975989	1.000441	24.8011	265.3219	252.6686	
P/BARNARD 3	1892	12	11.17460	1.432189	0.589631	31.2663	207.3318	170.0013	6.52
DENNING	1892	5	11.72750	1.970726	1.000371	89.6971	254.2404	129.3253	
BROOKS	1893	1	6.99490	1.195108	1.000000	143.8558	186.4458	85.2162	
BROOKS	1893	9	19.72230	0.811991	0.996489	129.8298	175.7157	347.4517	
RORDAME-QUENISSET	1893	7	7.77150	0.674549	0.999460	159.9742	338.1403	47.1231	
P/DENNING	1894	2	9.95400	1.147197	0.698400	5.5297	85.0784	46.3229	7.42
GALE	1894	4	13.88990	0.983032	0.989884	86.9721	207.1864	324.1878	958.00
P/SWIFT	1895	8	21.31560	1.297763	0.652013	2.9988	171.0633	167.7736	7.20
BROOKS	1895	10	21.55270	0.843034	1.000000	76.2460	83.8498	298.7764	
PERRINE	1895	12	18.82920	0.191978	1.000051	141.6209	321.2614	272.6723	
SWIFT	1896	4	18.14730	0.566286	1.000476	55.5665	179.0019	1.7393	
PERRINE-LAMP	1896	2	1.27660	0.587289	1.000000	155.7435	209.5886	358.3244	
SPERRA	1896	7	11.44280	1.142895	1.000000	88.4275	151.7416	41.0557	
P/GIACOBINI	1896	10	28.53420	1.454747	0.588483	11.3569	194.1883	140.5202	6.65

Comet	perihelion time (*)			q	e	i	Node	Peri	P
(P/ - periodic comets)	year	month	day	[AU]					[a]
PERRINE	1897	2	8.60400	1.062791	1.000983	146.1368	87.2233	172.3277	
PERRINE	1897	12	9.14260	1.356679	1.000000	69.6049	32.7908	65.9039	
BROOKS	1898	11	23.65240	0.756011	1.000000	140.3462	97.0380	123.5421	
PERRINE	1898	3	17.63200	1.095261	0.980454	72.5294	263.1675	47.3141	419.00
PERRINE	1898	8	16.70560	0.626438	1.000000	70.0306	259.8270	205.6204	
PERRINE-CHOFARDET	1898	10	21.03960	0.420373	1.000000	28.8547	35.6046	162.3672	
GIACOBINI	1898	7	25.99660	1.501318	1.000000	166.8527	278.9903	22.3753	
CODDINGTON-PAULY	1898	9	14.54530	1.701595	1.000934	69.9347	74.7125	233.2624	
CHASE	1898	9	20.61770	2.284591	0.999507	22.5059	96.5611	4.6251	
SWIFT	1899	4	13.47810	0.326576	1.000357	146.2634	25.7045	8.7040	
GIACOBINI	1899	9	15.40030	1.785838	1.000000	76.9432	272.9233	10.7800	
GIACOBINI	1900	4	29.41080	1.331529	1.001058	146.4441	41.0886	24.4010	
BORELLY-BROOKS	1900	8	3.70050	1.014835	1.000410	62.5284	328.7136	12.4257	
GREAT COMET(VISCARA)	1901	4	24.75130	0.244812	1.000000	131.0798	110.3297	203.0444	
BROOKS	1902	5	7.79600	0.443920	1.000000	65.1970	53.3370	229.6150	
PERRINE	1902	11	24.35670	0.401074	0.999968	156.3511	50.0299	152.9704	
BORRELLY	1903	8	28.10870	0.330182	1.000406	85.0123	294.2109	127.2578	
GIACOBINI	1903	3	16.50570	0.410561	0.999670	30.9351	2.9522	133.7093	
GIACOBINI	1903	3	23.97580	2.773525	1.000603	43.8991	118.1179	5.8237	
GRIGG	1903	3	25.92030	0.499420	1.000000	66.4890	213.7920	184.9460	
BROOKS	1904	3	7.63940	2.707793	1.001343	125.1292	276.4276	53.5356	
GIACOBINI	1904	11	3.80480	1.882049	1.000678	99.6036	219.0913	40.7453	
BROOKS	1905	12	22.79630	1.296297	1.000195	126.4402	287.0120	89.8570	
GIACOBINI	1905	4	4.54480	1.114641	0.969961	40.2052	158.0576	358.2080	226.00
KOPFF	1905	10	18.25450	3.339884	1.001510	4.2758	342.9054	158.6176	
SCHAER	1905	10	26.26260	1.052201	1.000000	140.5811	223.5568	132.7014	
THIELE	1906	11	21.72640	1.212686	0.982624	56.3892	85.4091	8.6320	583.00
ROSS	1906	2	21.43330	0.722913	1.000000	83.4601	72.6548	276.4576	
GIACOBINI	1906	1	22.85650	0.215900	1.000000	43.6529	92.6785	199.2087	
P/METCALF	1906	10	10.07940	1.631163	0.584480	14.6271	195.1690	199.7842	7.78
THIELE	1906	11	21.72640	1.212686	0.982624	56.3892	85.4091	8.6320	583.00
GIACOBINI	1907	3	19.60920	2.051705	1.001003	141.6610	97.7776	317.1094	
GRIGG-MELLISH	1907	3	28.05340	0.923861	1.000000	110.0635	189.7190	328.7579	
DANIEL	1907	9	4.46170	0.512173	0.998784	8.9633	143.5885	294.4510	
MELLISH	1907	9	14.85630	0.982949	1.000000	119.6315	55.1942	294.3723	

P/TEMPEL-SWIFT(JAVELLE)	1908	10	5.07490	1.152561	0.638120	5.4453	291.0609	113.4992	5.68
MOREHOUSE	1908	12	26.25710	0.945299	1.000672	140.1775	103.7536	171.5853	
BORRELLY-DANIEL	1909	6	5.75760	0.842844	0.994762	52.0760	306.1768	5.0017	
P/HALLEY(WOLF)	1910	4	20.17770	0.587189	0.967297	162.2151	57.8467	111.7170	76.10
GREAT JANUARY COMET	1910	1	17.58810	0.128975	0.999995	138.7817	89.3295	320.9024	
METCALF	1910	9	16.78000	1.948013	0.999797	121.0528	290.0849	50.9643	
BROOKS	1911	10	28.23680	0.489429	0.997005	33.8063	293.5000	153.0148	
KIESS	1911	6	30.67620	0.683837	0.996296	148.4265	157.9638	110.3678	
BELJAWSKI	1911	10	10.76380	0.303424	1.000147	96.4664	89.1976	71.7046	
QUENISSET	1911	11	12.84930	0.787623	0.998167	108.0951	35.7313	122.0314	
BORRELLY	1912	10	21.45840	1.107110	1.000000	124.6407	144.3151	99.6795	
GALE	1912	10	5.45310	0.716079	1.000451	79.8061	297.5479	25.6272.	
P/WESTPHAL(DELAVAN)	1913	11	26.82600	1.254020	0.919830	40.8840	347.3020	57.1120	61.90
METCALF	1913	9	14.55380	1.356120	0.997560	143.3610	157.9386	117.7028	
SCHAUMASSE	1913	5	15.59790	1.456831	0.995288	152.3622	315.5714	52.9664	
ZLATINSKI	1914	5	8.87290	0.543135	1.000000	112.9771	33.1559	116.3960	
KRITZINGER	1914	6	4.69380	1.198527	1.000106	23.9197	199.4116	72.2321	
NEUJMIN	1914	7	30.48660	3.746800	1.003290	71.0404	270.8130	14.0060	
CAMPBELL	1914	8	5.47080	0.712756	0.998666	77.8295	0.8845	270.3460	
DELAVAN	1914	10	26.76640	1.104459	1.000155	68.0361	59.7001	97.4607	
MELLISH	1915	7	17.65230	1.005286	1.000151	54.7908	72.7588	247.7767	
MELLISH	1915	10	13.89590	0.443290	1.000000	53.5450	78.2000	118.8400	
SCHAUMASSE	1917	5	18.71030	0.764184	1.000000	158.7320	10.1320	119.1560	
P/MELLISH	1917	4	11.17510	0.190186	0.993121	32.6831	87.9799	121.3070	145.00
WOLF	1917	6	17.07350	1.686446	0.999405	25.6656	183.7566	120.6239	
REID	1918	6	6.37940	1.101640	1.000000	69.7030	18.1410	194.9030	
P/SCHORR	1918	9	30.02340	1.882365	0.468443	5.5745	118.3354	279.0906	6.66
P/BRORSEN-METCALF	1919	10	17.36100	0.484280	0.972010	19.2100	311.1690	129.5690	72.00
METCALF	1919	12	7.80920	1.115291	1.000222	46.3862	121.4039	185.7461	
SKJELLERUP	1920	1	3.17400	0.297690	1.000000	123.1700	316.0200	276.5800	
SKJELLERUP	1920	12	11.50740	1.147892	0.994081	22.0329	108.1487	340.8716	
REID	1921	5	10.44960	1.008457	1.000357	132.1871	268.7468	64.4989	
REID	1921	10	28.76610	1.629083	0.986968	32.4444	274.8905	183.6932	
P/DUBIAGO	1921	5	5.34680	1.114978	0.929088	22.3245	66.5231	97.4325	62.30
BAADE	1922	10	26.52910	2.258774	1.000755	51.4610	220.8761	118.3040	
SKJELLERUP	1923	1	4.22320	0.923590	0.993735	23.3662	262.3430	264.4991	
DUBIAGO-BERNARD	1923	11	18.34810	0.777832	1.000000	113.7960	227.0416	255.3301	
REID	1924	3	13.94930	1.755695	0.998598	72.3306	114.3771	271.4341	
FINSLER	1924	9	4.84260	0.405993	1.000000	120.1490	80.4217	66.5319	
REID	1925	7	29.86690	1.633299	0.995116	26.9733	6.3441	259.2747	
ORKISZ	1925	4	1.50650	1.109477	1.000605	100.0172	318.4118	36.1854	
SHAJN-COMAS SOLA	1925	9	6.94640	4.180785	1.002432	146.7065	357.8403	205.7595	

Comet	perihelion time (*)			q	e	i	Node	Peri	P
(P/ - periodic comets)	year	month	day	[AU]					[a]
VAN BIESBROECK	1925	10	2.97300	1.566243	1.000382	49.3246	334.9126	106.4021	
WILK-PELTIER	1925	12	7.26740	0.763569	1.000505	144.6034	141.0781	126.1235	
REID	1926	12	30.41200	0.751640	1.000000	83.6657	108.9967	224.4317	
BLATHWAYT	1926	1	2.80420	1.345477	0.992384	128.3037	136.4246	328.2533	
ENSOR	1926	2	11.97280	0.322588	1.000000	122.9983	282.7623	354.8930	
P/NEUJMIN 2 (NEUJMIN)	1927	1	16.22830	1.338182	0.566836	10.6329	328.0036	193.7240	5.43
SKJELLERUP-MARISTANY	1927	12	18.18090	0.176157	0.999840	85.1117	77.5456	47.1523	
BLATHWAYT	1927	2	14.50300	1.035570	1.000000	92.3572	19.1164	231.8486	
STEARNS	1927	3	22.74910	3.683902	0.998179	87.6574	214.9517	11.1407	
PELTIER-SCHWASSMANN-WACHMANN	1930	1	15.60990	1.087114	0.998131	99.8888	147.8334	325.0805	
WILK	1930	1	22.30980	0.672235	0.999028	124.5168	179.2739	157.4949	
WILK	1930	3	28.79970	0.481810	0.992194	67.1416	90.5408	46.9528	485.00
BEYER	1930	4	18.11850	2.078632	1.000524	71.9727	116.6703	24.6212	
FORBES	1930	5	10.43760	1.152800	1.000000	97.0897	278.5668	320.9721	
NAGATA	1931	6	11.76750	1.046903	0.979203	42.3019	191.5395	320.1495	357.00
RYVES	1931	8	25.91250	0.074924	0.999326	169.2900	101.5460	168.1119	
CARRASCO	1931	11	30.59560	2.330615	1.000000	58.0573	18.0996	110.3024	
PELTIER-WHIPPLE	1932	9	1.85900	1.037198	0.976369	71.7130	344.7631	38.4725	291.00
HOUGHTON-ENSOR	1932	2	28.92810	1.254431	0.972127	74.2828	212.7838	303.5199	302.00
GEDDES	1932	9	21.07470	2.313581	1.001431	124.9936	215.3981	329.6995	
NEWMAN	1932	9	24.55050	1.647412	1.000602	78.3916	245.3898	69.7933	
DODWELL-FORBES	1932	12	30.53030	1.130788	0.972386	24.5018	77.9035	327.3463	262.00
PELTIER	1933	2	6.69900	1.000691	1.000000	86.6739	311.7675	135.9919	
JACKSON	1934	9	6.92050	3.485700	1.000000	141.9480	73.5040	124.3030	
JOHNSON	1935	2	26.46870	0.811148	0.991304	65.4259	91.7516	18.3898	901.00
VAN BIESBROECK	1936	5	11.63610	4.043409	1.002045	66.1084	299.8605	44.9016	
PELTIER	1936	7	8.95520	1.099868	0.991775	78.5497	134.2431	148.4711	
KAHO-KOZIK-LIS	1936	7	15.82030	0.518403	0.994389	121.9417	264.3117	45.8573	888.00
HUBBLE	1936	11	14.24560	1.953657	0.972499	11.5820	97.1283	147.4607	599.00
FINSLER	1937	8	15.66580	0.862744	0.999985	146.4127	58.7134	114.8260	
P/WILK	1937	2	21.53410	0.618937	0.981098	26.0175	57.5714	31.4619	187.00
WHIPPLE	1937	6	20.06240	1.733791	1.000137	41.5560	127.9151	107.7273	
P/GALE(CUNNINGHAM)	1938	6	18.48320	1.182907	0.760719	11.7261	67.2549	209.1266	11.00
P/HERSCHEL-RIGOLLET	1939	8	9.46400	0.748490	0.974050	64.2000	355.2820	29.2980	155.00
KOZIK-PELTIER	1939	2	6.85350	0.716496	0.995103	63.5211	288.9120	169.0309	

JURLOF-ACHMAROF-HASSEL	1939	4	10.16880	0.528266	0.998477	138.1164	311.5836	89.2464	
FRIEND	1939	11	5.62600	0.945209	0.997192	92.9581	196.4134	126.7793	
WHIPPLE-PARASKEVOPOULOS	1940	10	8.24910	1.082228	0.980843	54.6955	134.3661	235.7316	425.00
OKABAYASI-HONDA	1940	8	15.75920	1.061768	1.001459	133.1185	127.3655	329.6716	
FRIEND-REESE-HONDA	1941	1	20.37780	0.941864	0.981221	26.2697	329.0854	132.7383	355.00
CUNNINGHAM	1941	1	16.23400	0.367751	1.000485	49.8909	295.8873	199.5763	
DE KOCK-PARASKEVOPOULOS	1941	1	27.65770	0.790033	0.999102	168.1995	42.3853	268.6762	
VAN GENT	1941	9	3.18420	0.874789	1.000243	94.5174	256.8618	85.3284	
WHIPPLE-BERNASCONI-KULIN	1942	4	30.83330	1.445303	1.000893	79.4457	340.2358	223.4174	
P/VAISAELAE 2	1942	2	15.81950	1.287080	0.933640	38.0026	171.5925	335.2223	85.40
OTERMA	1942	9	27.27830	4.113405	1.003183	172.5127	280.3886	163.6688	
WHIPPLE-FEDTKE-TEVZADZE	1943	2	6.72110	1.353647	0.992196	19.7144	100.1180	39.8245	
DAIMACA	1943	8	21.54650	0.758300	1.000000	161.3180	82.6690	36.3930	
VAN GENT	1944	7	17.61330	2.225933	1.002085	95.0113	202.8020	336.9772	
VAN GENT-PELTIER-DAIMACA	1944	1	12.25000	0.874300	1.000000	136.1780	57.9260	33.0700	
FRIEND-PELTIER	1945	12	17.28100	0.194350	1.000000	49.4740	325.5020	216.7160	
VAISAELAE	1945	1	4.06220	2.410856	0.993500	17.2827	28.3978	238.9305	
DU TOIT	1945	5	17.14560	0.998063	1.000000	156.5092	254.3616	280.1397	
DU TOIT	1945	12	27.96520	0.007516	1.000000	141.8669	350.5027	72.0627	
TIMMERS	1946	4	13.26320	1.724129	1.001169	72.8472	128.9679	54.3231	
PAJDUSAKOVA-ROTBART-WEBER	1946	5	11.43660	1.018252	1.000000	169.5580	301.3064	22.2617	
JONES	1946	10	26.77910	1.136113	1.000776	56.9677	237.6333	320.4200	
BESTER	1947	2	7.36340	2.407663	1.000947	108.1692	34.8588	348.6206	
BESTER	1947	5	31.11310	1.402863	1.000000	111.4129	232.3665	357.3948	
BECVAR	1947	5	3.91980	0.961836	1.000000	129.1500	322.3800	182.1300	
RONDANINA-BESTER	1947	5	20.96280	0.559799	0.997427	39.2949	353.2111	303.7548	
WIRTANEN	1947	7	18.34520	2.827968	1.001052	97.3266	311.0981	9.3797	
WIRTANEN	1947	9	4.43100	3.261102	1.002278	155.0818	121.4108	73.4524	
HONDA	1947	11	17.90600	0.753090	1.000000	106.2860	311.9620	221.2410	
SOUTHERN COMET	1947	12	2.58450	0.110032	0.999548	138.5357	336.6077	196.1552	
JOHNSON	1948	4	8.95380	4.708380	1.000000	53.2378	139.7156	191.8346	
PAJDUSAKOVA-MRKOS	1948	5	16.61230	2.107056	1.000787	92.9214	246.9543	66.9043	
BESTER	1948	2	16.42330	0.748125	1.000363	140.5674	270.7491	350.2248	
BESTER	1948	10	22.86600	1.273428	0.997499	87.6034	66.9702	274.2030	
MRKOS	1948	2	16.69230	1.499557	1.001085	77.5391	198.6019	61.9268	
HONDA-BERNASCONI	1948	5	15.90520	0.207628	0.999875	23.1547	203.1152	317.0634	
ECLIPSE COMET	1948	10	27.42710	0.135421	0.999935	23.1223	210.3321	107.2617	
WIRTANEN	1949	5	1.10910	2.517207	0.999352	130.2713	119.9002	229.9433	
WILSON-HARRINGTON	1949	10	11.81500	1.032930	1.000000	3.9010	272.4980	87.5960	
BAPPU-BOK-NEWKIRK	1949	10	26.54910	2.058177	0.998644	105.7641	309.0127	89.5348	
JOHNSON	1950	1	19.31310	2.553231	1.000713	131.3555	221.6327	40.0968	
PAJDUSAKOVA	1951	1	30.42630	0.719063	1.000000	87.8921	310.5211	68.6065	

Comet	perihelion time (*)			q	e	i	Node	Peri	P
(P/ - periodic comets)	year	month	day	[AU]					[a]
MINKOWSKI	1951	1	15.04290	2.572335	1.001231	144.1504	38.1857	192.4608	
P/HARRINGTON-WILSON	1951	10	30.41450	1.664094	0.515277	16.3542	127.7894	342.9961	6.36
PELTIER	1952	7	15.25730	1.201978	0.999915	45.5624	187.9426	96.5892	
MRKOS	1952	6	8.67220	1.282517	0.981763	112.0223	121.6538	144.8555	590.00
WILSON-HARRINGTON	1952	1	12.94960	0.740427	0.999739	152.5327	76.1833	269.6066	
MRKOS	1953	1	24.85990	0.777729	1.000343	97.1773	342.8832	253.8232	
HARRINGTON	1953	1	5.41770	1.664977	0.995910	59.1199	220.6904	191.6314	
MRKOS-HONDA	1953	5	26.43580	1.022132	0.997391	93.8561	275.1874	85.7464	
P/PONS-BROOKS(ROEMER)	1954	5	22.88030	0.773656	0.954846	74.1784	255.1918	199.0340	70.90
PAJDUSAKOVA	1954	1	24.67140	0.072347	1.000000	13.5761	114.5577	94.0537	
HARRINGTON	1954	1	18.09970	2.060608	1.000000	136.6415	292.0390	1.4262	
ABELL	1954	3	24.11520	4.495680	1.002791	123.9273	320.6386	73.7544	
ABELL	1954	7	7.34920	0.970116	1.000564	53.2224	2.3350	194.3817	
VAZAROVA	1954	6	1.93500	0.677461	1.000286	116.1581	122.1237	357.2544	
KRESAK-PELTIER	1954	8	29.66610	0.746377	1.000174	88.5395	74.8700	254.7211	
BAADE	1955	8	13.30500	3.869934	1.000509	100.3892	264.6425	144.6773	
HONDA	1955	8	4.06030	0.884616	1.000947	107.5159	338.7233	348.1948	
MRKOS	1955	6	4.16690	0.534419	0.989361	86.4978	48.2434	32.5090	356.00
BAKHAREV-MACFARLANE-KRIENKE	1955	7	11.54400	1.427450	0.994167	50.0289	302.7630	13.2250	
P/OLBERS(MRKOS)	1956	6	19.13920	1.178450	0.930334	44.6118	85.4121	64.6356	69.60
P/CROMMELIN(PAJDUS.-HENDRIE)	1956	10	25.20580	0.743360	0.919170	28.8811	250.4291	195.9769	27.90
MRKOS	1956	4	13.60920	0.842240	1.000000	147.4560	226.0253	81.0110	
HARO-CHAVIRA	1956	1	26.66450	4.076871	1.004667	79.5985	72.1749	57.2911	
WIRTANEN	1957	9	2.41370	4.447262	1.002745	33.2052	232.9493	13.2695	
MRKOS	1957	8	1.43730	0.354933	0.999365	93.9392	67.6252	40.3135	
AREND-ROLAND	1957	4	8.03240	0.316035	1.000168	119.9493	215.1596	308.7852	
LATYSHEV-WILD-BURNHAM	1957	12	5.12930	0.539130	1.000000	156.7210	210.1860	277.6230	
P/OTERMA	1958	6	10.50080	3.387823	0.144488	3.9922	155.1093	354.8724	7.88
BURNHAM	1958	4	16.30360	1.322689	0.999943	15.7938	150.5054	16.4559	
BESTER-HOFFMEISTER	1959	7	17.55560	1.250534	1.002679	12.8344	105.0633	186.4793	
MRKOS	1959	11	13.21620	1.253464	0.999748	19.6356	99.9396	84.6737	
BURNHAM	1959	9	27.72750	1.170751	1.000000	69.5008	83.0833	273.2409	
BURNHAM-SLAUGHTER	1959	3	11.51980	1.628198	0.999866	61.2520	323.0783	100.7401	
ALCOCK	1959	8	17.60750	1.150354	1.000919	48.2628	159.2257	124.7062	
ALCOCK	1959	9	15.88030	0.165507	1.000000	107.7663	225.0714	300.6112	

HUMASON	1959	12	11.20500	4.266927	1.000884	125.4652	306.5669	46.4667	
P/SCHAUMASSE(JEFFERS-GIBSON)	1960	4	18.05500	1.195943	0.705435	12.0187	86.2405	51.9462	8.18
BURNHAM	1960	3	20.99560	0.504353	1.000145	159.6017	251.9572	306.6525	
WILSON-HUBBARD	1961	7	17.49660	0.040173	0.999905	24.2085	298.2442	270.7084	
P/VAN HOUTEN	1961	4	26.84280	3.956877	0.366820	6.6541	22.9286	14.3911	15.60
CANDY	1961	2	8.91750	1.061612	0.989900	150.9618	176.5797	136.6144	
SEKI	1961	10	10.64080	0.681129	0.991812	155.7128	246.6704	126.5900	759.00
HONDA	1962	4	20.19860	0.653093	1.000000	72.8728	79.1217	72.1344	
HUMASON	1962	12	10.20500	2.133412	0.989569	153.2841	154.7355	233.5502	
SEKI-LINES	1962	4	1.66290	0.031397	1.000003	65.0101	303.9767	11.4787	
ALCOCK	1963	5	5.88160	1.537101	0.998060	86.2151	42.7665	146.6173	
IKEYA	1963	3	21.47480	0.632139	0.993377	160.6452	52.5045	336.2851	932.00
PEREYRA	1963	8	23.95640	0.005065	0.999946	144.5757	7.2388	86.1576	903.00
ANDERSON	1963	10	9.29000	2.087400	1.000000	4.6700	99.6200	343.1100	
IKEYA	1964	8	1.21110	0.821752	0.984643	171.9195	269.2965	290.8080	391.00
TOMITA-GERBER-HONDA	1964	6	30.62160	0.500363	0.995933	161.8277	309.1857	58.4394	
EVERHART	1964	8	23.18630	1.259275	0.996513	67.9672	279.7440	20.6924	
P/TEMPEL-TUTTLE(SCHUBART-BES.)	1965	4	30.00100	0.981656	0.904433	162.7106	234.4357	172.5795	32.90
P/DE VICO-SWIFT(KLEMOLA)	1965	8	23.24450	1.624298	0.524494	3.6064	24.4216	325.3516	6.31
ALCOCK	1965	10	26.14020	1.293900	1.000000	65.0157	174.9249	150.5239	
IKEYA-SEKI	1965	10	21.18370	0.007786	0.999915	141.8577	346.2974	69.0502	880.00
P/NEUJMIN 1 (ANDREWS)	1966	12	9.40400	1.542568	0.774824	15.0255	347.1842	346.8107	17.90
IKEYA-EVERHART	1966	8	5.32100	0.881880	1.000000	48.2810	106.9670	50.0390	
BARBON	1966	4	17.79140	2.018766	0.998183	28.7123	166.9923	136.1638	
KILSTON	1966	10	28.07240	2.384748	0.999376	40.2709	155.3556	154.4802	
MITCHELL-JONES-GERBER	1967	6	16.78510	0.178251	1.000000	56.7030	31.5974	78.9394	
SEKI	1967	3	13.74250	0.457263	1.000000	106.4815	199.8424	144.3161	
RUDNICKI	1967	1	20.88430	0.419225	1.000391	9.0787	75.0494	79.7433	
WILD	1967	3	2.52860	1.327158	1.000000	99.1015	306.1471	173.2659	
P/PERRINE-MRKOS(TAMMANN)	1968	11	1.53740	1.272212	0.642630	17.7619	240.1590	166.0641	6.72
HONDA	1968	8	7.99310	1.160434	1.000665	149.2408	108.0404	88.7049	
HONDA	1968	11	3.83810	1.098991	1.000000	127.9141	252.6003	282.8153	
IKEYA-SEKI	1968	2	25.70550	1.696581	0.999152	129.3165	254.6278	70.8724	
WILD	1968	3	31.25200	2.609070	1.000000	135.1710	208.5020	101.6930	
TAGO-HONDA-YAMAMOTO	1968	5	16.27590	0.680374	0.995583	102.1693	232.4077	50.4557	
WHITAKER-THOMAS	1968	6	4.50310	1.234003	1.000000	61.7700	254.0145	353.9833	
BALLY-CLAYTON	1968	8	20.84480	1.771085	1.000000	93.1642	318.6971	26.8683	
THOMAS	1969	1	12.24730	3.316033	0.995301	45.2230	15.4138	82.5765	
FUJIKAWA	1969	10	12.44800	0.773917	1.000000	8.9319	191.6874	299.0297	
TAGO-SATO-KOSAKA	1969	12	21.26770	0.472638	0.999920	75.8196	100.9629	267.8274	
P/DU TOIT-NEUJMIN-DELPORTE	1970	10	7.20860	1.672228	0.509736	2.8481	189.2289	114.2645	6.30
DAIDO-FUJIKAWA	1970	2	15.80530	0.065742	1.000000	100.1706	29.9065	266.6497	

comet	perihelion time (*)			q	e	i	Node	Peri	P
(P/ - periodic comets)	year	month	day	[AU]					[a]
BENNETT	1970	3	20.04460	0.537606	0.996193	90.0437	223.9589	354.1510	
KOHOUTEK	1970	3	21.63740	1.719085	0.999125	86.3089	301.0594	123.4739	
WHITE-ORTIZ-BOLELLI	1970	5	14.48590	0.008879	1.000000	139.0652	336.3186	61.2935	
SUZUKI-SATO-SEKI	1970	10	1.79540	0.405499	1.000000	60.7818	292.9785	318.5286	
ABE	1970	10	20.69900	1.112539	1.000048	126.7141	21.0007	96.5799	
GEHRELS	1971	1	6.53410	3.276561	0.996943	175.6103	24.0084	128.6899	
TOBA	1971	4	17.24970	1.233292	1.000775	109.6846	103.3753	152.3562	
P/NEUJMIN 3 (ROEMER-MC CALL.)	1972	5	16.57200	1.976325	0.589705	3.8867	150.2054	146.9271	10.60
BRADFIELD	1972	3	27.70050	0.927214	0.998126	123.6993	159.5991	257.6958	
HECK-SAUSE	1972	10	5.46390	2.511124	1.000501	138.6348	175.1787	346.2352	
SANDAGE	1972	11	14.80930	4.275707	1.006288	79.3723	224.7868	56.6628	
ARAYA	1972	12	18.94480	4.860748	0.999910	113.0852	314.1872	267.2089	
P/WILD 1 (ROEMER-LATTA)	1973	7	2.84230	1.980511	0.647087	19.8906	358.2012	167.9464	13.30
KOHOUTEK	1973	6	7.18140	1.382019	0.998723	121.6046	164.1185	74.8584	
KOHOUTEK	1973	12	28.43070	0.142425	1.000008	14.3051	257.7656	37.8238	
SANDAGE	1973	11	8.14810	4.812056	1.000059	137.4017	278.5459	72.5523	
P/GEHRELS 1	1973	1	24.74230	2.935319	0.506783	9.6444	14.6286	28.9452	14.50
KOJIMA	1973	2	12.27750	2.147032	1.001018	141.8510	42.4838	334.2980	
HUCHRA	1973	3	11.58570	2.384384	0.972800	48.3239	57.1492	123.4850	821.00
GIBSON	1973	8	9.88100	3.842070	1.000643	108.0689	243.9041	221.2626	
P/SCHWASSMANN-WACHMANN 1	1974	2	15.32950	5.447856	0.105428	9.7469	319.6364	14.4681	15.00
P/DU TOIT(TORRES)	1974	4	1.49690	1.294036	0.786982	18.7026	22.1402	257.2073	15.00
P/WIRTANEN(ROEMER-VAUGHN)	1974	7	5.66550	1.256059	0.614209	12.2677	83.5425	351.8301	5.87
BENNETT	1974	12	1.52290	0.864646	1.000000	134.8232	50.6446	324.9597	
BRADFIELD	1974	3	18.35630	0.503191	0.999697	61.2898	143.0379	333.1302	
CESCO	1974	5	12.96310	1.372826	0.978898	173.1648	165.0304	176.7406	525.00
VAN DEN BERGH	1974	8	7.77490	6.018928	1.003948	60.8578	225.4027	151.7654	
P/AREND(ROEMER-DANIEL)	1975	5	24.68360	1.846905	0.537640	19.9556	355.6584	46.9222	7.98
SUZUKI-SAIGUSA-MORI	1975	10	15.36020	0.838047	0.985653	118.2381	216.1091	152.0241	446.00
BRADFIELD	1975	4	4.57930	1.216939	1.001410	55.2500	157.2120	264.1272	
BRADFIELD	1975	12	21.18130	0.218719	1.000001	70.6259	270.6124	358.0972	
P/BOETHIN	1975	1	5.63130	1.093694	0.779544	5.9117	26.9698	11.1276	11.00
SCHUSTER	1975	1	15.55010	6.880784	1.002153	112.0202	22.0874	193.4247	
P/SMIRNOVA-CHERNYKH	1975	8	6.25450	3.567170	0.145385	6.6414	77.1079	90.1810	8.53
LOVAS	1975	8	22.18130	3.011456	0.999602	50.6422	11.6714	261.3640	

KOBAYASHI-BERGER-MILON	1975	9	5.33480	0.425561	1.000095	80.7779	295.6526	116.9756	
MORI-SATO-FUJIKAWA	1975	12	25.87750	1.603933	0.997450	91.6061	277.9800	246.2475	
P/PONS-WINNECKE(ROEMER-HELLER)	1976	11	28.74160	1.254209	0.634704	22.2943	92.7442	172.3613	6.36
P/WOLF(ROEMER-VAUGHN)	1976	1	25.35730	2.500775	0.395684	27.3319	203.8087	161.1453	8.42
P/HARRINGTON-ABELL(ROE.-DAN.)	1976	4	21.83820	1.776310	0.540040	10.1689	336.7758	138.4528	7.59
P/KLEMOLA(SAUSE)	1976	8	10.19370	1.765580	0.641717	10.6401	181.5717	148.8816	10.90
P/CHURYUMOV-GERASIMENKO	1976	4	7.21980	1.298504	0.630562	7.1251	50.3748	11.3127	6.59
BRADFIELD	1976	2	24.63860	0.847810	0.993811	46.8403	160.0926	313.0044	
BRADFIELD	1976	2	25.06100	0.678320	1.000000	147.7760	69.4950	221.7550	
LOVAS	1976	7	6.38720	5.857415	1.003931	86.6298	285.3383	118.7960	
LOVAS	1976	10	31.98250	5.715207	1.003506	64.5204	337.4712	146.2772	
SATO	1976	1	3.92120	0.863968	1.001214	93.9534	280.7962	215.4724	
WEST	1976	2	25.22160	0.196626	0.999971	43.0700	118.2313	358.4190	
HARLAN	1976	11	3.15010	1.568877	0.999695	38.8059	80.7206	193.2540	
P/FAYE(ROEMER-HELLER)	1977	2	27.83180	1.609767	0.575613	9.0910	199.0854	203.6587	7.39
P/KOPFF(ROEMER-HELLER)	1977	3	7.97580	1.571988	0.545361	4.7245	120.3323	162.9080	6.43
P/TAYLOR(KOWAL)	1977	1	11.45980	1.951491	0.465384	20.5564	108.2146	355.5819	6.97
P/JOHNSON(ROEMER-HELLER)	1977	1	8.42820	2.195661	0.386061	13.9066	117.7948	206.2092	6.76
P/GEHRELS 3	1977	4	23.21890	3.424339	0.151873	1.1013	242.5502	231.4802	8.11
WEST	1977	7	21.55940	5.606362	1.002964	116.9353	210.9270	343.2887	
P/KOWAL 1	1977	2	22.55680	4.663515	0.237034	4.3586	28.4357	178.0493	15.10
HELIN	1977	6	30.97290	1.117568	1.000000	43.1893	19.4429	249.3525	
TSUCHINSHAN	1977	7	24.80810	3.603034	0.999419	168.5432	4.5810	318.1617	
P/SANGUIN	1977	9	17.55620	1.810591	0.663913	18.6373	182.2660	162.0830	12.50
P/TRITTON	1977	10	28.62270	1.438523	0.580318	7.0348	300.0140	147.7233	6.35
KOHLER	1977	11	10.56990	0.990575	0.999546	48.7182	181.8234	163.4877	
P/TEMPEL 1 (SHAO-SCHW.-MC CRO)	1978	1	11.01450	1.496853	0.519501	10.5449	68.3398	179.0780	5.50
P/TUTTLE-GIACOBINI-KRESAK	1978	12	25.71010	1.123755	0.642927	9.9410	153.3262	49.3758	5.58
P/DENNING-FUJIKAWA	1978	10	2.04200	0.778570	0.820170	8.6680	40.9680	334.0450	9.01
P/WOLF-HARRINGTON(SCHW.-S.-S.)	1978	3	15.89760	1.614585	0.538013	18.4588	254.2133	186.9817	6.53
P/DANIEL(STANDEN)	1978	7	8.43770	1.661793	0.549893	20.1366	68.4927	10.8328	7.09
P/COMAS SOLA(MC CR.-SCHW.-S.)	1978	9	24.26970	1.869866	0.565821	12.9582	62.4280	42.8370	8.94
P/WHIPPLE(PEREYRA-LABORDE)	1978	3	27.53600	2.468633	0.352245	10.2461	188.3387	189.9764	7.44
P/JACKSON-NEUJMIN(KOWAL)	1978	12	25.44100	1.425350	0.654330	14.0740	163.1330	196.2810	8.37
P/ASHBROOK-JACKSON(PER.-MOYAN)	1978	8	19.88020	2.284203	0.399858	12.5252	2.0589	349.0139	7.43
P/AREND-RIGAUX(MC CROSKY-PER.)	1978	2	2.49210	1.441665	0.599554	17.8557	121.5241	328.9891	6.83
P/VAN BIESBR.(MC CROSKY)	1978	12	3.02870	2.395473	0.552513	6.6186	148.5782	134.2519	12.40
P/WILD 2	1978	6	15.87740	1.491085	0.556713	3.2751	136.1147	39.8767	6.17
P/TSUCHINSHAN 1 (BULGER-SC.-S)	1978	5	7.45600	1.499472	0.576111	10.5044	96.2054	22.7793	6.65
P/TSUCHINSHAN 2 (SEKI)	1978	9	20.99540	1.784568	0.504148	6.7149	287.6390	203.1811	6.83
P/KOJIMA(KOSAI-HURUKAWA)	1978	5	24.83190	2.398874	0.392891	0.8855	154.0831	348.5580	7.85
BRADFIELD	1978	3	17.69060	0.436585	1.000000	51.0869	259.7942	48.7131	

Comet (P/ - periodic comets)	perihelion time (*) year	month	day	q [AU]	e	i	Node	Peri	P [a]
BRADFIELD	1978	9	29.08810	0.431870	1.000000	138.2584	357.7202	240.4498	
P/CLARK(SEKI)	1978	11	26.28730	1.557066	0.500937	9.4994	59.0804	209.0452	5.51
P/SCHUSTER	1978	1	6.81660	1.627544	0.574039	20.4438	50.8371	353.9162	7.47
P/CHERNYKH	1978	2	14.94180	2.568400	0.594386	5.7279	134.1153	266.7035	15.90
MACHOLZ	1978	8	13.67320	1.771588	1.000416	130.6336	289.9845	224.7618	
SEARGENT	1978	9	14.84870	0.369872	0.998269	67.8197	41.0821	207.7518	
P/HANEDA-CAMPOS	1978	10	9.49610	1.101403	0.665170	5.9506	131.5862	240.4361	5.97
MEIER	1978	11	11.40930	1.136527	1.000844	43.7550	348.6456	231.4030	
P/GICLAS	1978	11	21.25040	1.731562	0.511885	8.5432	141.5345	247.2500	6.68
P/HOLMES(SHAO-SCHWARTZ)	1979	2	22.65980	2.160003	0.413115	19.2046	327.3804	23.6005	7.06
P/GIACOBINI-ZINNER(SHAO-SCH.)	1979	2	12.77580	0.996045	0.714655	31.7000	195.0698	171.9734	6.52
P/SCHWASSMANN-WACHMANN 2	1979	3	17.03840	2.135006	0.386707	3.7298	125.9424	357.4594	6.50
P/SCHWASSMANN-WACHMANN 3(JO.)	1979	9	2.78380	0.940867	0.693664	11.4076	69.2744	198.7429	5.38
P/SHAJN-SCHALDACH(SHAO-SCHW.)	1979	1	9.02260	2.223258	0.406733	6.1542	167.1953	215.2950	7.25
BRADFIELD	1979	7	23.26260	0.413217	1.000059	136.2357	163.5165	47.6495	
BRADFIELD	1979	12	21.60030	0.545183	0.987645	148.6040	102.5099	257.5760	293.00
P/KOWAL 2	1979	1	13.86310	1.522149	0.567364	15.8097	247.1171	189.4718	6.60
MEIER	1979	10	17.34580	1.432081	0.973197	67.0806	296.9298	112.5787	391.00
P/RUSSELL 1	1979	5	26.97920	1.613440	0.516822	22.6531	230.1765	0.2293	6.10
TORRES	1979	7	15.40260	4.686916	1.001060	92.1752	292.4382	10.1068	
HOOWARD-KOOMEN-MICHELS	1979	8	30.91900	0.001640	1.000000	142.6760	350.1030	72.0650	
P/ENCKE	1980	12	6.57680	0.339939	0.846763	11.9460	334.1956	185.9827	3.30
P/TUTTLE(SHAO-SCHWARTZ)	1980	12	14.70540	1.014937	0.822575	54.4622	269.8811	206.8924	13.70
P/WEST-KOHOUTEK-IKEMURA(SCHU.)	1980	4	12.45480	1.400759	0.581428	30.0662	84.6087	358.1260	6.12
P/STEPHAN-OTERMA (SCHUSTER)	1980	12	5.16750	1.574360	0.860000	17.9809	78.5122	358.1610	37.70
P/BROOKS 2 (SCHUSTER)	1980	11	25.39260	1.849711	0.489786	5.5464	176.2361	198.2213	6.90
P/BORRELLY(SCHUSTER)	1980	2	20.02960	1.319246	0.631488	30.2008	75.0577	352.7716	6.77
P/REINMUTH 1 (SCHWARTZ-SHAO)	1980	10	29.74200	1.981564	0.487103	8.3060	121.1063	9.4624	7.59
P/REINMUTH 2 (JEKABSONS)	1980	1	29.94950	1.945857	0.454910	6.9692	296.0492	45.4068	6.74
P/FORBES(SCHUSTER)	1980	9	25.32000	1.478871	0.564961	4.6656	23.0013	262.5588	6.27
P/HONDA-MRKOS-PAJDUSAKOVA	1980	4	11.07300	0.580608	0.808623	13.1184	232.9286	184.6326	5.28
P/HARRINGTON(JEKABSONS)	1980	12	24.01880	1.604534	0.555719	8.6516	118.9527	233.0384	6.86
P/WILD 3	1980	10	5.18040	2.287518	0.368027	15.4612	72.0479	179.3317	6.89
BRADFIELD	1980	12	29.54170	0.259823	0.999725	138.5882	114.6465	358.2855	
P/KOHOUTEK(SCHUSTER)	1980	4	17.66910	1.570647	0.536631	5.4163	273.1213	169.9183	6.24

MEIER	1980	12	9.65000	1.519550	0.994657	100.9806	24.7375	87.9642	
P/RUSSELL 2	1980	5	19.53680	2.160579	0.416249	12.5313	44.4534	245.4411	7.12
TORRES	1980	4	19.87410	2.583929	1.000000	73.1449	278.8228	334.9775	
CERNIS-PETRAUSKAS	1980	6	22.44090	0.523249	1.000000	49.0739	159.9282	337.8154	
P/LOVAS	1980	9	3.44000	1.675682	0.614520	12.2928	342.3256	72.5699	9.06
PANTHER	1980	1	27.32300	1.657271	0.998974	82.6419	331.2992	105.6034	
RUSSELL	1980	3	6.34280	2.110704	1.000000	128.7049	232.0736	297.2715	
BOWELL	1980	3	12.31200	3.363815	1.057196	1.6651	114.0676	134.8791	
P/SWIFT-GEHRELS(SHAO-SCHWARTZ)	1981	11	27.26690	1.361128	0.691291	9.2422	314.0339	84.5408	9.26
P/FINLAY(JEKABSONS-CANDY)	1981	6	20.00080	1.100927	0.698421	3.6425	41.8014	322.1293	6.97
P/VAISAELAE 1 (GIBSON)	1981	7	30.62910	1.799860	0.633417	11.6069	134.5116	47.9336	10.90
P/SLAUGHTER-BURNHAM(SCHW.-SH.)	1981	11	18.90490	2.544226	0.503950	8.1531	345.9324	44.1667	11.60
P/KEARNS-KWEE(SHEFFER)	1981	11	30.54160	2.223542	0.485531	8.9855	315.2621	131.3808	8.99
P/GEHRELS 2 (COCHRAN)	1981	11	18.66740	2.361641	0.408480	6.6627	215.5340	183.4639	7.98
P/LONGMORE(SEKI)	1981	10	21.56520	2.399919	0.342922	24.4131	15.0088	195.9250	6.98
GONZALES	1981	3	25.66160	2.333601	0.999395	107.1522	143.2700	181.6112	
P/HOWELL	1981	5	4.35820	1.615399	0.507452	5.5685	75.3536	214.6089	5.94
P/BUS	1981	6	11.35640	2.182572	0.374744	2.5783	181.5278	24.6446	6.52
BUS	1981	7	30.70160	2.458484	1.000000	160.6597	23.5536	189.8163	
ELIAS	1981	8	18.21300	4.742492	1.000596	115.3169	176.0068	310.2393	
P/TEMPEL 2 (GIBSON)	1982	6	1.53550	1.381404	0.544893	12.4375	119.1579	190.9220	5.29
P/D´ARREST(GIBSON-EVERHART)	1982	9	14.31140	1.291086	0.624811	19.4301	138.8598	176.9682	6.38
P/GRIGG-SKJELLERUP(GIBSON)	1982	5	14.99470	0.989245	0.665681	21.1366	212.6324	359.3282	5.09
P/DU TOIT-HARTLEY	1982	3	30.44500	1.194664	0.601961	2.9385	308.5845	251.6724	5.20
P/GUNN	1982	11	26.86180	2.459149	0.316361	10.3832	67.8849	196.9829	6.82

(*) - before 1940: Universaltime ; 1940 onward: Ephemeris Time

2. REFERENCE MODEL OF THE EARTH CORE

radius	density	v_P	v_S	pressure	gravity
[10^3 m]	[10^3 kg/m^3]	[10^3 m/s]	[10^3 m/s]	[10^8 Pa]	[10^2 m/s^2]
0	13.08840	11.26220	3.66700	3638.524	0.00
100	13.08630	11.26064	3.66670	3636.131	36.56
200	13.07977	11.25593	3.66342	3628.956	73.11
300	13.06888	11.24809	3.65794	3617.011	109.61
400	13.05364	11.23712	3.65027	3600.315	146.04
500	13.03404	11.22301	3.64041	3578.894	182.39
600	13.01009	11.20576	3.62835	3552.783	218.62
700	12.98000	11.18000	3.61400	3522.024	254.73
800	12.94000	11.16000	3.59700	3486.665	290.68
900	12.91211	13.52100	3.57905	3446.764	326.45
1000	12.87073	11.10542	3.55823	3402.383	362.03
1100	12.82501	11.07249	3.53522	3353.596	397.39
1200	12.77493	11.03643	3.51002	3300.480	432.51
1221	12.76360	11.02827	3.50432	3288.513	440.02
1221	12.16634	10.35568	0.00000	3288.502	440.03
1300	12.12500	10.30971	0.00000	3245.423	463.68
1400	12.06924	10.24959	0.00000	3187.493	494.13
1500	12.00989	10.18743	0.00000	3126.159	524.77
1600	11.94682	10.12291	0.00000	3061.461	555.48
1700	11.87990	10.05572	0.00000	2993.457	586.14
1800	11.80900	9.98554	0.00000	2922.221	616.69
1900	11.73401	9.91206	0.00000	2847.839	647.04
2000	11.65478	9.83496	0.00000	2770.407	677.15
2100	11.57119	9.75393	0.00000	2690.035	706.97
2200	11.48311	9.66865	0.00000	2606.838	736.45
2300	11.39042	9.57881	0.00000	2520.942	765.56
2400	11.29298	9.48409	0.00000	2432.484	794.25
2500	11.19067	9.38418	0.00000	2341.603	822.48
2600	11.08335	9.27876	0.00000	2248.453	850.23
2700	10.97091	9.16752	0.00000	2153.189	877.46
2800	10.85321	9.05015	0.00000	2055.978	904.14
2900	10.73012	8.92632	0.00000	1956.991	930.23
3000	10.60152	8.79573	0.00000	1856.479	955.70
3100	10.46727	8.65805	0.00000	1754.418	980.51
3200	10.32726	8.51298	0.00000	1651.209	1004.64
3300	10.18134	8.36019	0.00000	1546.982	1028.04
3400	10.02940	8.19939	0.00000	1441.941	1050.65
3480	9.90349	8.06482	0.00000	1357.510	1068.23
3480	5.56645	13.71660	7.26466	1357.509	1068.23
3500	5.55641	13.71168	7.26486	1345.619	1065.32
3600	5.50642	13.68753	7.26575	1287.067	1052.04
3630	5.49145	13.68041	7.26597	1269.742	1048.44
3630	5.49145	13.68041	7.26597	1269.741	1048.44
3700	5.45657	13.59597	7.23403	1229.719	1040.66
3800	5.40681	13.47742	7.18892	1173.465	1030.95
3900	5.35706	13.36074	7.14423	1118.207	1022.72

v_P - longitudinal seismic velocity

v_S - trannsversal seismic velocity

3.1. Formulary

3.1.1. Nomenclature

B — magnetic field
c_s — sound velocity
g — gravitational acceleration
l — characteristic length scale
m — mass
M — Ion/proton mass ratio, $M = m_i/m_p$
n — number density
p — pressure
T — temperature
v — characteristic flow velocity
v_A — Alfvenic velocity
Z — charge state
γ — ratio of specific heats
ε — permittivity
η — magnetic viscosity, $\eta = 1/(\mu\sigma)$
$\varkappa$ — thermal diffusivity
λ — collisional mean free path
μ — permeability
υ — kinematic viscosity
ϱ — mass density
σ — electrical conductivity
Ω — rotational velocity

Convention for 3.1.4.—3.1.6.:
All number densities are in cm^{-3}, lengths in m, velocities in km/s, temperatures in K, magnetic fields in nT.

3.1.2. Physical constants

Boltzmann's constant	$k_B = 1.380662 * 10^{-23}$ J K^{-1}
Universal gas constant	$R = 8.31441$ J K^{-1} mol^{-1}
Gravitational constant	$G = 6.6720 * 10^{-11}$ N m^2 kg^{-2}
Astronomical unit	1 AU = $149.6 * 10^6$ km
Velocity of light in vacuo	$c = 2.9979246 * 10^8$ m s^{-1}
Elementary charge	$e = 1.6021892 * 10^{-19}$ A s
Electron mass	$m_e = 9.109534 * 10^{-31}$ kg
Proton mass	$m_p = 1.6726485 * 10^{-27}$ kg
Proton/electron mass ratio	$m_p/m_e = 1836.15152$
Permittivity of free space	$\varepsilon_0 = 8.85418782 * 10^{-12}$ F m^{-1}
Permeability of free space	$\mu_0 = 1.2566370614 * 10^{-6}$ H m^{-1}

3.1.3. Dimensionless numbers of magnetohydrodynamics

Alfvenic Mach number $$M_A = \frac{v}{v_A}$$

Chandrasekhar number $$Q = Ha^2 = St * Re = \frac{B^2 l^2}{\mu\varrho\eta\nu}$$

Ekman number $$Ek = \frac{Ro}{Re} = \frac{\upsilon}{l^2\Omega}$$

Euler number $$Eu = \frac{p}{\varrho v^2}$$

Froude number $\mathrm{Fr} = \frac{\Omega^2 l}{g}$

Hall number $H = \frac{eBl}{vm}$

Hartmann number $\mathrm{Ha} = \frac{Bl}{(\mu\varrho\eta\nu)^{1/2}}$

Knudsen number $\mathrm{Kn} = \frac{\lambda}{l}$

Magnetic Prandtl number $P_m = \frac{R_m}{\mathrm{Re}} = \frac{\upsilon}{\eta}$

Magnetic Reynolds number $R_m = \frac{lv}{\eta}$

Plasma beta $\beta = \frac{2\mu\varrho}{B^2}$

Prandtl number $\mathrm{Pr} = \frac{\upsilon}{\varkappa}$

Reynolds number $\mathrm{Re} = \frac{lv}{\upsilon}$

Rossby number $\mathrm{Ro} = \frac{v}{\Omega l}$

Sonic mach number $M_s = \frac{v}{c_s}$

Stewart number $\mathrm{St} = \frac{B^2 l}{\mu\varrho\eta\nu}$

3.1.4. Frequencies

Electron plasma frequency $f_{pe} = \frac{\omega_{pe}}{2\pi} = 8.98 * 10^3 (n_e)^{1/2}$ [Hz]

$$\omega_{pe} = \left(\frac{n_e e^2}{\varepsilon_0 m_e}\right)^{1/2} = 5.64 * 10^4 (n_e)^{1/2} \text{ [Hz]}$$

Ion plasma frequency $f_{pi} = \frac{\omega_{pi}}{2\pi} = 2.1 * 10^2 \left(\frac{n_i Z^2}{M}\right)^{1/2}$ [Hz]

$$\omega_{pi} = \left(\frac{n_i e^2 Z^2}{m_i}\right)^{1/2} = 1.32 * 10^3 \left(\frac{n_i Z^2}{M}\right)^{1/2} \text{ [Hz]}$$

Electron gyrofrequency $f_{ce} = \frac{\omega_{ce}}{2\pi} = 28\,B$ [Hz]

$$\omega_{ce} = \frac{eB}{m_e} = 1.76 * 10^2 B \text{ [Hz]}$$

Ion gyrofrequency

$$f_{ci} = \frac{\omega_{ci}}{2\pi} = 1.53 * 10^{-2} \frac{BZ}{M} \text{ [Hz]}$$

$$\omega_{ci} = \frac{ZeB}{m_i} = 9.59 * 10^{-2} \frac{BZ}{M} \text{ [Hz]}$$

Upper hybrid frequency

$$f_{uH} = \frac{\omega_{uH}}{2\pi}$$

$$\omega_{uH} = (\omega_{pe}^2 + \omega_{ce}^2)^{1/2}$$

Lower hybrid frequency

$$f_{lH} = \frac{\omega_{lH}}{2\pi} = 0.654\, B \left(\frac{Z}{M}\right)^{1/2} \text{ [Hz]}$$

$$\omega_{lH} = (\omega_{ci} \cdot \omega_{ce})^{1/2} = 4.10\, B \left(\frac{Z}{M}\right)^{1/2} \text{ [Hz]}$$

3.1.5. Velocities

Electron thermal velocity

$$v_{tc} = \left(\frac{2k_B T_e}{m_e}\right)^{1/2} = 5.51\,(T_e)^{1/2} \text{ [km/s]}$$

Ion thermal velocity

$$v_{ti} = \left(\frac{2k_B T_i}{m_i}\right)^{1/2} = 0.129 \left(\frac{T_i}{M}\right)^{1/2} \text{ [km/s]}$$

Ion sound velocity

$$c_{si} = \left(\frac{2Zk_B T_e}{m_i}\right)^{1/2} = 0.129 \left(\frac{ZT_e}{M}\right)^{1/2} \text{ [km/s]}$$

Sound velocity (perfect gas)

$$c_s = \left(\frac{\gamma p}{\varrho}\right)^{1/2}$$

Alfven velocity

$$v_A = \frac{B}{(\mu n_i m_i)^{1/2}} = 21.8 \frac{B}{(n_i M)^{1/2}} \text{ [km/s]}$$

3.1.6. Lengths

Landau length

$$l_L = \frac{e^2}{4\pi\varepsilon_0 k_B T_e} = 1.67 * 10^{-5} T_e^{-1} \text{ [m]}$$

Debye length

$$l_D = \left(\frac{\varepsilon_0 k_B T_e}{e^2 n_e}\right)^{1/2} = 6.90 * 10^{-2} \left(\frac{T_e}{n_e}\right)^{1/2} \text{ [m]}$$

Electron gyroradius

$$r_e = \frac{v_{te}}{\omega_{ce}} = 31.3 \frac{(T_e)^{1/2}}{B} \text{ [m]}$$

Ion gyroradius

$$r_i = \frac{v_{ti}}{\omega_{ci}} = 1.35 * 10^3 \frac{(MT_i)^{1/2}}{ZB} \text{ [m]}$$

3.2. Solar wind

Distance from Sun AU	r/r_{Sun}	n_{sw} [cm^{-3}]	v_{sw} [$km s^{-1}$]	T_{esw} [10^4 K]	T_{psw} [10^4 K]	$\|B_{sw}\|$ [nT]	$<\|B_R\|>$ [nT]	$<\|B_T\|>$ [nT]	$<\|B_N\|>$ [nT]
0.4	86.0	59.300 [3]	404.0 [3]	22.0 [3]	22.00 [3]				
0.4	86.0	26.200 [21]/1/	494.0 [21]/1/		47.00 [21]/1/	28.00 [21]/1/			
0.4	86.0	139.600 [21]/1/	359.0 [21]/1/		7.70 [21]/1/	17.60 [21]/1/			
0.6	129.0	26.100 [3]	408.0 [3]	19.0 [3]	14.00 [3]				
0.6	129.0	10.400 [21]/1/	618.0 [21]/1/		27.80 [21]/1/	17.20 [21]/1/			
0.6	129.0	20.600 [21]/1/	377.0 [21]/1/		6.50 [21]/1/	12.70 [21]/1/			
0.8	172.0	14.600 [3]	409.0 [3]	17.0 [3]	9.60 [3]				
0.8	172.0	5.600 [21]/1/	667.0 [21]/1/		32.00 [21]/1/	9.40 [21]/1/			
0.8	172.0					4.70 [21]/1/			
1.0	215.0	2.100 [21]/1/	717.0 [21]/1/		30.50 [21]/1/	6.00 [1]	2.9 [18]/2/	4.2 [18]/2/	2.9 [18]/2/
1.0	215.0	9.350 [3]	410.0 [3]	16.0 [3]	7.00 [3]	6.00 [13]	2.7 [19]/4/	3.3 [19]/4/	1.9 [19]/4/
1.0	215.0	7.000 [13]	430.0 [13]	15.0 [13]	7.00 [13]	6.70 [18]/2/			
1.0	215.0	8.300 [20]/6/	401.5 [18]/2/			5.70 [19]/4/			
1.0	215.0		420.0 [20]/6/						
1.5	322.5	4.320 [2]	294.0 [2]	9.8 [2]	4.00 [3]	3.80 [18]/2/	1.6 [18]/2/	2.4 [18]/2/	1.7 [18]/2/
1.5	322.5	4.150 [3]	410.0 [3]	14.0 [3]	5.60 [26]/2/	5.30 [19]/4/	1.5 [19]/4/	2.0 [19]/4/	1.2 [19]/4/
1.5	322.5		403.2 [18]/2/			3.81 [27]/4/			
2.0	430.0	2.390 [2]	297.0 [2]	7.8 [2]	2.70 [3]	2.60 [18]/2/	1.0 [18]/2/	1.7 [18]/2/	1.1 [18]/2/
2.0	430.0	2.330 [3]	411.0 [3]	13.0 [3]	5.60 [26]/2/	5.10 [19]/4/	0.9 [19]/4/	1.4 [19]/4/	0.8 [19]/4/
2.0	430.0		404.3 [18]/2/			2.66 [27]/4/			
4.0	860.0	0.590 [2]	303.0 [2]	4.2 [2]		1.22 [27]/4/			
4.0	860.0	0.580 [3]	411.0 [3]	9.9 [3]	1.00 [3]	4.60 [19]/4/	0.3 [19]/4/	0.6 [19]/4/	0.4 [19]/4/
4.0	860.0		407.1 [18]/2/		4.10 [26]/2/	1.00 [18]/2/	0.4 [18]/2/	0.7 [18]/2/	0.4 [18]/2/

Distance from Sun AU	r/r_{Sun}	n_{sw} [cm^{-3}]	v_{sw} [kms^{-1}]	T_{esw} [10^4 K]	T_{psw} [10^4 K]	$\lvert B_{sw}\rvert$ [nT]	$\langle\lvert B_R\rvert\rangle$ [nT]	$\langle\lvert B_T\rvert\rangle$ [nT]	$\langle\lvert B_N\rvert\rangle$ [nT]
6.0	1290.0	0.260 [2]	305.0 [2]	2.7 [2]	0.57 [3][3	0.80 [27]/4/			
6.0	1290.0	0.260 [3]	411.0 [3]	8.6 [3]		4.30 [19]/4/	0.2 [19]/4/	0.4 [19]/4/	0.2 [19]/4/
6.0	1290.0		408.8 [18]/2/		2.93 [26]/2/	0.60 [18]/2/	0.2 [18]/2/	0.4 [18]/2/	0.2 [18]/2/
8.0	1720.0	0.150 [2]	306.0 [2]	1.9 [2]	2.66 [26]/2/	0.60 [27]/4/			
8.0	1720.0	0.150 [3]	411.0 [3]	7.7 [3]	0.38 [3]				
10.0	2150.0	93 [2]	307.0 [2]	1.5 [2]	2.03 [26]/2/	0.48 [27]/4/			
10.0	2150.0	93 [3]	411.0 [3]	7.2 [3]	0.28 [3]				

r - Distance from Sun

r_{Sun} - Sun radius

N_{sw} - Particle number density

V_{sw} - Velocity

T_{esw} - Electron temperature

T_{psw} - Proton temperature

$\lvert B_{sw}\rvert$ - Magnetic field magnitude

$\langle\lvert B_R\rvert\rangle$ - Averaged radial component for the magnetic field

$\langle\lvert B_T\rvert\rangle$ - Avaraged tangential component of the magnetic field

$\langle\lvert B_N\rvert\rangle$ -Avaraged normal component of the magnetic field

Body	Bow shock distance in r_{body} from body centre		Iono/magneto-pause distance in r_{body} from body centre		magnetic flux density at the surface	magnetic dipole moment	ratio dipole: quadrupole: octupole moment	inclination of dipole axis *)
	parallel	vertical	parallel	vertical	[nT]	$[Tm^3]$		[degree]
Mercury	1.4 [5] / 7/		1.4 [12]		400 [50] / 7/	2.3E12 [50] / 7/	1:0.45:0.29 [34] / 7/	7.0 [50] / 7/
Mercury					50 [46] / 7/	5E12 [46] / 7/		
Mercury						2.4E12 [34] / 7/		
Mercury						1.8E12 [33] / 7/		
Mercury						6E12 [5] / 7/		
Venus	1.13 [4] / 8/	2.15 [4] / 8/	1.19 [11]	1.13 [28]	<0.4 [40] / 9/	<8.4E10 [40] / 9/		
Venus	1.3 [9] / 8/	2.04 [9] / 8/	1.0 [12]					
Venus	1.57 [10] / 9/	2.36 [10] / 9/						
Earth	13.5 [4] /10/	26.1 [4] /10/	10 [11]		3.1E4 [37]	8E15 [37]	1:0.14:0.1 [37]	11.5 [37]
Earth	13.9 [4] /11/	26.5 [4] /11/	10 [12]					
Earth	14.2 [4] /12/	26.5 [4] /12]						
Earth	13.4 [4] /13/	25.9 [4] /13/						
Moon					10 [48] /14/	5E10 [48] /14/		

Body	Bow shock distance in r_{body} from body centre		Iono/magneto-pause distance in r_{body} from body centre		magnetic flux density at the surface	magnetic dipole moment	ratio dipole: quadrupole: octupole moment	inclination of dipole axis *)
	parallel	vertical	parallel	vertical	[nT]	[Tm^3]		[grad]
Mars	1.5 [4] /15/	2.4 [4] /15/	1.5 [11]		60 [39] /15/	2.1E12 [38] /16/		15 [42]
Mars			1 [12]			2.4E12 [39] /15/		
Mars						5E11 [36] /17/		
Mars						4.5E14 [41]		
Jupiter	72.0 [14] / 4/	124.0 [14] / 4/	56.9 [11]	100.0 [14] / 4/	1.2E6 [44] / 4/	1.4E20 [42] / 2/	1:0.25:0.20 [42] / 3/	10.8 [42] / 3/
Jupiter	84.0 [14] / 5/	116.0 [14] / 5/	100.0 [12]	85.0 [14] / 5/		1.55E20 [43] / 3/		
Jupiter			56.0 [14] / 4/			1.5E20 [44] / 4/		
Jupiter			68.0 [14] / 5/					
Io			1.15 [17]		3000 [49]	2E13 [49]		
Europa			1.15 [17]		1200 [49]	4E12 [49]		
Ganymede			1.16 [17]		300 [49]	6E12 [49]		
Callisto			1.15 [17]		100 [49]	1E12 [49]		

Body	Bow shock distance in r_{body} from body centre parallel	Bow shock distance in r_{body} from body centre vertical	Iono/magneto-pause distance in r_{body} from body centre parallel	Iono/magneto-pause distance in r_{body} from body centre vertical	magnetic flux density at the surface [nT]	magnetic dipole moment [Tm^3]	ratio dipole: quadrupole: octupole moment	inclination of dipole axis *) [grad]
Saturn	27.0 [15] / 4/	35.0 [15] / 4/	48.0 [11]	25.0 [15] / 4/	2.0E4 [45] / 3/	4.7E18 [45] / 3/	1:0.07 [45] / 3/	0.0 [15] / 4/
Saturn	27.0 [16]		20.0 [12]			4.7E18 [15] / 4/		
Saturn			23 [15] /4/					
Saturn			20 [16]					
Titan	no bow [29] / 4/	shock	1.75 [30]	1.75 [30]	100 [49]	2E12 [49]		
Titan						no i.f. [29] / 4/		
Uranus	22.5 [57] / 5/	58.8 [57] / 5/	17.8 [57] / 5/	33.2 [57] / 5/	1.0E4 [57] / 5/	3.86E17 [57] / 5/		60 [57] / 5/
Uranus		46.2 [57] / 5/		25.0 [57] / 5/	1.1E5 [57] / 5/			
Neptune					2.8E4 [47]			
Pluto					1.4E3 [47]			
Halley	1.6E6 [54] /19/	7.6E5 [54] /19/	4.7E3 [55] /19/	3.8E3 [55] /19/				
Halley		4.5E5 [56] /20/						
Halley	1.1E6 [53] /18/	5.0E5 [53] /18/						

parallel - parallel to the axis Sun-body   vertical - vertical to the axis Sun-body

no i.f. - no intrinsic field

*) - to the rotation axis

For Uranus:

1. The dipole is shifted -0.31 r_{body} in the direction of the planetary angular momentum vector.
2. Due to the large inclination of the dipole axis the vertical distance of bow shock and magnetopause varies in time between the given values.
3. Due to the fact,that the rotation axis is in the orbital plane,the surface flux density in sunward direction(upper value) is much lower than in antisunward direction(lower value).

For Halley:

parallel - inbound measurement along the spacecraft trajectory in [km]

vertical - outbound measurement along the spacecraft trajectory in [km]

3.4. Venus

3.4.1. Models for lower atmosphere, below cloud top for Venus [22]

	Latitudes below 40°			60° Latitude		
Z [km]	T [K]	p [MPa]	ρ [kgm^{-3}]	T [K]	p [MPa]	ρ [kgm^{-3}]
0	733	9.2100	65.000	737	9.2100	64.600
2	717	8.1100	58.700	722	8.1200	58.400
4	701	7.1200	52.900	706	7.1300	52.600
6	685	6.2400	47.600	691	6.2500	47.200
8	669	5.4400	42.500	675	5.4600	42.300
10	653	4.7300	37.900	659	4.7500	37.700
12	637	4.1000	33.800	644	4.1200	33.600
14	623	3.5400	29.800	627	3.5600	29.900
16	609	3.0500	26.300	611	3.0700	26.400
18	595	2.6200	23.200	594	2.6300	23.300
20	579	2.2400	20.400	577	2.2500	20.500
22	562	1.9000	17.800	561	1.9100	17.900
24	545	1.6100	15.500	544	1.6100	15.600
26	529	1.3550	13.480	527	1.3560	13.560
28	512	1.1350	11.670	509	1.1340	11.720
30	495	0.9450	10.050	493	0.9430	10.070
32	478	0.7820	8.610	477	0.7800	8.600
34	461	0.6430	7.340	461	0.6400	7.300
36	446	0.5250	6.190	445	0.5220	6.170
38	430	0.4250	5.200	429	0.4230	5.180
40	416	0.3420	4.320	415	0.3400	4.310
42	403	0.2730	3.560	400	0.2710	3.560
44	391	0.2170	2.910	388	0.2150	2.910
46	379	0.1710	2.370	375	0.1690	2.360
48	366	0.1336	1.910	359	0.1316	1.920
50	349	0.1034	1.553	344	0.1014	1.545
52	332	0.0791	1.248	324	0.0772	1.248
54	312	0.0595	0.999	303	0.0577	0.995
56	291	0.0440	0.792	282	0.0422	0.783
58	274	0.0318	0.608	262	0.0302	0.602
60	263	0.0227	0.451	243	0.0210	0.452
62	252	0.0159	0.330		0.0000	0.000
64	244	0.0110	0.236		0.0000	0.000
66	236	0.0076	0.168		0.0000	0.000

Z - Height (relative to a Venus radius of 6052.0)

3.4.2. Models for the middle atmosphere, 60 to 100 km for Venus [22]

			0° to 30° Latitude			45° Latitude			60° Latitude			75° Latitude			85° Latitude		
Z	g	R	T	p	ρ	T	p	ρ	T	p	ρ	T	p	ρ	T	p	ρ
[km]	[ms^{-2}]	[*)]	[K]	[hPa]	[kgm^{-3}]	[K]	[hPa]	[kgm^{-3}]	[K]	[hPA]	[kgm^{-3}]	[K]	[hPa]	[kgm^{-3}]	[K]	[hPa]	[kgm^{-3}]
60	8.696	191.40	263	227.0000	0.451000	258	221.0000	0.447000	243	210.0000	0.452000	235	190.0000	0.422000	243	173.0000	0.372000
62	8.690	191.40	252	159.0000	0.330000	247	154.0000	0.326000	228	143.0000	0.327000	213	127.0000	0.311000	243	119.0000	0.256000
64	8.684	191.40	244	110.0000	0.236000	240	106.0000	0.231000	217	95.0000	0.229000	220	82.9000	0.197000	243	81.9000	0.176000
66	8.679	191.40	236	76.0000	0.168000	235	73.0000	0.161000	219	62.6000	0.149000	232	55.5000	0.125000	243	56.4000	0.121000
68	8.673	191.40	237	51.8000	0.114000	240	49.5000	0.108000	240	42.2000	0.091800	238	37.7000	0.082900	242	38.8000	0.083800
70	8.667	191.40	233	35.2000	0.078900	236	33.8000	0.074900	237	28.9000	0.063600	239	25.8000	0.056400	242	26.7000	0.057600
72	8.662	191.40	228	23.8000	0.054500	231	23.0000	0.051900	236	19.7000	0.043600	238	17.7000	0.038800	241	18.3000	0.039800
74	8.656	191.40	223	15.9000	0.037300	226	15.4000	0.035700	234	13.4000	0.029900	235	12.0000	0.026800	238	12.6000	0.027600
76	8.650	191.40	217	10.5400	0.025400	221	10.3000	0.024400	227	9.0600	0.020800	230	8.1600	0.018500	235	8.5800	0.019100
78	8.645	191.40	211	6.9100	0.017100	214	6.8000	0.016600	218	6.0400	0.014500	224	5.4800	0.012800	230	5.8200	0.013200
80	8.639	191.40	203	4.4700	0.011500	207	4.4300	0.011200	210	3.9600	0.009850	217	3.5700	0.008600	224	3.9100	0.009110
82	8.634	191.40	194	2.8400	0.007650	198	2.8400	0.007490	202	2.5600	0.006610	206	2.3300	0.005910	215	2.5900	0.006290
84	8.628	191.44	184	1.7600	0.005000	189	1.7800	0.004920	194	1.6200	0.004300	195	1.4900	0.003980	203	1.6800	0.004330
86	8.622	191.46	176	1.0700	0.003170	181	1.0900	0.003160	185	1.0100	0.002840	187	0.9270	0.002590	192	1.0700	0.002900
88	8.617	191.53	171	0.6370	0.001950	175	0.6600	0.001970	177	0.6130	0.001810	180	0.5680	0.001650	183	0.6600	0.001880
90	8.611	191.57	167	0.3740	0.001170	169	0.3910	0.001210	170	0.3650	0.001120	174	0.3420	0.001020	176	0.4000	0.001190
92	8.605	191.66	164	0.2170	0.000690	165	0.2280	0.000722	165	0.2130	0.000675	169	0.2020	0.000625	170	0.2380	0.000730
94	8.600	191.80	162	0.1250	0.000402	163	0.1320	0.000423	163	0.1230	0.000395	166	0.1180	0.000372	168	0.1400	0.000434
96	8.594	191.93	162	0.0719	0.000231	164	0.0764	0.000243	165	0.0715	0.000226	166	0.0690	0.000217	168	0.0820	0.000254
98	8.589	192.06	163	0.0415	0.000132	167	0.0445	0.000139	173	0.0421	0.000127	168	0.0404	0.000125	170	0.0483	0.000148
100	8.583	192.24	170	0.0242	0.000074	173	0.0263	0.000079	177	0.0253	0.000074	173	0.0239	0.000072	173	0.0287	0.000086

Z - Height (relative to a Venus radius of 6052.0 km)

*) - [$Jkg^{-1} K^{-1}$]

3.4.3. Model for the upper atmosphere, 100 to 180 km for Venus [22]

	Antisolar or midnight model SZA >120°				Subsolar or noon model SZA <50°				
Z	T	p	ρ	R	T	p	ρ	R	g
[km]	[K]	[hPa]	[kgm^{-3}]	[$Jkg^{-1}K^{-1}$]	[K]	[hPa]	[kgm^{-3}]	[$Jkg^{-1}K^{-1}$]	[ms^{-2}]
100	170	2.50E- 2	7.64E- 5	192.2	170	2.50E- 2	7.64E- 5	192.2	8.583
104	160	8.48E- 3	2.75E- 5	192.6	172	8.81E- 3	2.66E- 5	192.6	8.572
108	150	2.70E- 3	9.31E- 6	193.1	175	3.16E- 3	9.36E- 6	193.1	8.561
112	140	7.95E- 4	2.93E- 6	193.6	179	1.16E- 3	3.36E- 6	193.6	8.550
116	129	2.15E- 4	8.56E- 7	194.3	184	4.41E- 4	1.23E- 6	194.3	8.539
120	117	5.17E- 5	2.26E- 7	195.3	190	1.73E- 4	4.66E- 7	195.1	8.527
124	125	1.23E- 5	4.99E- 8	197.1	196	7.00E- 5	1.82E- 7	196.3	8.517
128	135	3.30E- 6	1.22E- 8	200.6	204	2.95E- 5	7.30E- 8	198.3	8.506
132	142	9.91E- 7	3.35E- 9	208.2	212	1.31E- 5	3.05E- 8	202.1	8.494
136	146	3.30E- 7	1.02E- 9	220.1	220	6.07E- 6	1.33E- 8	208.1	8.483
140	147	1.26E- 7	3.26E-10	263.0	230	2.98E- 6	6.01E- 9	215.4	8.472
144	140	5.65E- 8	1.25E-10	324.0	240	1.55E- 6	2.87E- 9	224.4	8.461
148	134	2.83E- 8	5.41E-11	390.0	250	8.51E- 7	1.43E- 9	238.0	8.450
152	131	1.53E- 8	2.63E-11	442.0	260	4.96E- 7	7.54E-10	253.0	8.440
156	129	8.68E- 9	1.42E-11	475.0	272	3.06E- 7	4.16E-10	270.0	8.429
160	127	5.07E- 9	7.91E-12	504.0	289	1.99E- 7	2.35E-10	293.0	8.418
164	126	3.03E- 9	4.57E-12	526.0	299	1.37E- 7	1.44E-10	319.0	8.407
168	126	1.84E- 9	2.65E-12	551.0	300	9.78E- 8	9.42E-11	346.0	8.397
172	125	1.13E- 9	1.55E-12	581.0	300	7.16E- 8	6.43E-11	371.0	8.386
176	125	7.19E-10	9.33E-13	616.0	300	5.35E- 8	4.48E-11	398.0	8.375
180	125	4.73E-10	5.69E-13	665.0	300	4.08E- 8	3.22E-11	422.0	8.364

Z - height (relative to a Venus radius of 6052.0 km)
SZA - Solar zenit angle

3.4.4 Composition and particle number density of the ionosphere, measuered by /9/

height	neutral particle number density n_o from [6] [cm^{-3}]				ion number density n_i from [6] [cm^{-3}]				temperature from [7] [K]	
r/r_{Ven}	CO_2	CO	O	N_2	CO_2^+	O_2^+	O^+	C^+	T_e	T_i
1.012	2.2E10	3.0E9	2.1E9	1.2E9	1.8E4	5.6E5	8.0E2			
1.013	9.0E 8	3.8E8	6.2E8	1.5E8					340	240
1.015	4.8E 7	5.9E7	2.1E8	2.4E7					540	410
1.017					4.2E2	1.5E4	5.1E4	3.0E3	540	2440
1.021					1.9E2	4.6E2	2.6E4	2.3E3	1160	2830

r_{Ven} - Venus radius

3.5. Earth

3.5.1. Molecular composition of Earth ionosphere [32]

	Day-time (12:00 noon), concentration n [cm^{-3}]							Night-time (midnight), concentration n [cm^{-3}]						
Z [km]	N_2	O_2	He	O	H	total n	T [K]	N_2	O_2	He	O	H	total n	T [K]
60	55E14	15E14				70E14	270	55E14	15E14				70E14	270
70	16E14	42E13				20E14	200	16E14	42E13				20E14	200
80	23E13	62E12				29E13	180	23E13	62E12				29E13	180
90	31E12	82E11				39E12	190	31E12	82E11				39E12	190
100	77E11	19E11	57E 6	20E10	62E 3	96E11	210	77E11	19E11	57E 6	22E10	62E 3	96E11	210
110	14E11	35E10	38E 6	14E10	53E 3	19E11	270	14E11	35E10	38E 6	14E10	53E 3	19E11	270
120	58E10	12E10	25E 6	76E 9	43E 3	78E10	360	58E10	12E10	25E 6	76E 9	43E 3	78E10	360
130	20E10	38E 9	17E 6	37E 9	32E 3	28E10	460	20E10	37E 9	17E 6	37E 9	32E 3	27E10	470
150	48E 9	72E 8	10E 6	14E 9	21E 3	69E 9	670	48E 9	70E 8	10E 6	14E 9	22E 3	69E 9	650
200	48E 8	59E 7	49E 5	30E 8	13E 3	84E 8	1070	47E 8	56E 7	59E 5	32E 8	16E 3	84E 8	850
250	11E 8	12E 7	34E 7	12E 8	10E 3	25E 8	1250	77E 7	71E 6	43E 5	12E 8	14E 3	20E 8	910
300	32E 7	27E 6	27E 5	59E 7	93E 2	93E 7	1330	14E 7	10E 6	33E 5	44E 7	13E 3	59E 7	930
400	35E 6	22E 5	19E 5	16E 7	83E 2	20E 7	1390	60E 5	28E 4	21E 5	70E 6	11E 3	78E 6	940
500	44E 5	21E 4	14E 5	50E 6	76E 2	56E 6	1400	28E 4	86E 2	13E 5	12E 6	10E 3	13E 6	950
600	59E 4	21E 3	10E 5	16E 6	71E 2	17E 6	1400	15E 3	29E 1	87E 4	23E 5	90E 2	32E 5	950
700	85E 3	23E 2	80E 4	52E 5	66E 2	60E 5	1400	84E 1	11E 0	58E 4	44E 4	81E 2	10E 5	950
800	13E 3	27E 1	61E 4	18E 5	62E 2	24E 5	1400	51E-1	46E-2	39E 4	89E 3	73E 2	48E 4	950
900	21E 2	33E 0	47E 4	62E 4	58E 2	11E 5	1400	34E-1	21E-3	26E 4	19E 3	67E 2	28E 4	950
1000	35E 1	43E-1	37E 4	22E 4	54E 2	60E 4	1400	24E-2	10E-4	18E 4	42E 2	61E 2	19E 4	950

Z - Height

3.5.2. Structure of Earth ionosphere (electrons and ions) [32]

	Day-time					Night-time					Relative ion concentration n_i/n						
Z	n	T_e	T_i	ω_{pe}	ω_{pi}	n	T_e	T_i	ω_{pe}	ω_{pi}	NO^+	O_2^+	O^+	He^+	H^+	N^+	N_2^+
[km]	[cm^{-3}]	[K]	[K]	[Hz]	[Hz]	[cm^{-3}]	[K]	[K]	[Hz]	[Hz]							
60	80E00	270	270	50E04	21E02						0.800	0.200					
70	20E01	200	200	80E04	33E02						0.800	0.200					
80	10E02	180	180	18E05	75E02	10E00	180	180	18E04	75E01	0.700	0.300					
90	80E02	200	190	50E05	21E03	60E00	190	190	44E04	18E02	0.650	0.350					
100	80E03	240	210	16E06	67E03	12E02	210	210	19E05	80E02	0.620	0.380					
110	12E04	320	270	19E06	80E03	18E02	270	270	24E05	10E03	0.550	0.450					
120	13E04	400	360	20E06	84E03	21E02	360	360	26E05	11E03	0.390	0.600	0.01				
130	15E04	500	460	22E06	92E03	22E02	480	470	27E05	11E03	0.420	0.550	0.02				0.010
150	30E04	800	670	31E06	14E04	24E02	670	650	28E05	12E03	0.450	0.410	0.13				0.010
200	50E04	1300	1100	40E06	22E04	30E02	900	850	31E05	17E03	0.045	0.045	0.90			0.005	0.005
250	10E05	1700	1300	56E06	33E04	10E03	1000	910	56E05	33E03	0.006	0.005	0.98			0.006	0.003
300	16E05	2000	1400	71E06	42E04	10E04	1200	930	18E06	11E04			0.99			0.010	
400	15E05	2400	1450	69E06	41E04	30E04	1400	950	31E06	18E04			0.97	0.005	0.01	0.020	
500	90E04	2600	1600	54E06	32E04	20E04	1500	1000	25E06	15E04			0.90	0.015	0.03	0.060	
600	40E04	2700	2100	36E06	22E04	13E04	1600	1020	20E06	12E04			0.84	0.020	0.06	0.080	
700	20E04	2800	2200	25E06	16E04	80E03	1700	1100	16E06	10E04			0.75	0.040	0.11	0.100	
800	10E04	2870	2300	18E06	13E04	50E03	1800	1200	13E06	90E03			0.61	0.060	0.21	0.120	
900	70E03	2940	2400	15E06	12E04	30E03	1900	1300	98E05	77E03			0.41	0.090	0.40	0.100	
1000	50E04	3000	2500	13E06	12E04	20E03	2000	1400	80E05	73E03			0.28	0.140	0.51	0.070	

Z - Height

3.5.3. Effective electron collision frequencies in the Earth ionosphere [32]

Day-time effective collosions frequencies

Z	ν_{eN_2}	ν_{eO_2}	ν_{eHe}	ν_{eO}	ν_{eH}	ν_{em}	ν_{ei}	l_{ei}
[km]	[Hz]	[Hz	Hz]	[Hz]	[Hz]	[Hz]	[Hz]	[m]
60	3.1E 7	6.1E 6				3.7E 7	9.5E-1	1.1E5
70	7.0E 6	1.3E 6				8.3E 6	3.4E 0	2.6E4
80	9.2E 5	2.0E 5				1.1E 6	1.9E 1	4.4E3
90	1.3E 5	2.7E 4				1.6E 5	1.2E 2	7.3E2
100	4.0E 4	7.2E 3	4.1E-1	1.0E 3	4.4E-4	4.8E 4	8.4E 2	1.1E2
110	9.6E 3	1.6E 3	3.1E-1	7.0E 2	4.3E-4	1.2E 4	8.5E 2	1.3E2
120	4.9E 3	6.5E 2	2.5E-1	4.2E 2	4.1E-4	6.2E 3	5.8E 2	2.4E2
130	2.2E 3	2.4E 2	1.9E-1	2.5E 2	3.6E-4	2.7E 3	4.4E 2	3.5E2
150	7.1E 2	8.4E 1	1.4E-1	1.1E 2	2.9E-4	9.1E 2	4.8E 3	3.9E2
200	1.1E 2	9.6E 0	8.5E-2	3.1E-1	2.2E-4	1.5E 2	4.4E 2	5.3E2
250	3.0E 1	2.3E 0	6.5E-2	1.5E 1	1.9E-4	4.7E 1	6.5E 2	4.0E2
300	1.0E 1	5.9E-1	5.6E-2	7.3E 0	1.9E-4	1.8E 1	8.1E 2	3.4E2
400	1.3E 0	5.5E-2	4.3E-2	2.2E 0	1.8E-4	3.5E 0	5.9E 2	5.2E2
500	1.7E-1	5.5E-3	3.2E-2	6.9E-1	1.7E-4	9.0E-1	3.0E 2	1.8E3
600	2.4E-2	5.6E-4	2.4E-2	2.3E-1	1.6E-4	2.7E-1	1.4E 2	2.3E3
700	3.5E-3	6.4E-5	1.9E-2	7.5E-2	1.5E-4	9.8E-2	6.9E 1	4.8E3
800	5.5E-4	7.8E-6	1.5E-2	2.7E-2	1.5E-4	4.2E-2	3.4E 1	9.8E3
900	9.0E-5	9.8E-7	1.2E-2	1.2E-2	1.4E-4	2.4E-2	2.4E 1	1.4E4
1000	1.5E-5	1.3E-7	9.4E-3	3.3E-3	1.3E-4	1.3E-2	1.7E 1	2.0E4

Night-time effective collosions frequencies

Z	ν_{eN_2}	ν_{eO_2}	ν_{eHe}	ν_{eO}	ν_{eH}	ν_{em}	ν_{ei}	l_{ei}
[km]	[Hz]	[Hz	Hz]	[Hz]	[Hz]	[Hz]	[Hz]	[m]
80	9.2E 5	2.0E 5				1.1E 6	2.3E-1	3.6E5
90	1.3E 5	2.7E 4				1.6E 5	1.1E 0	7.4E4
100	3.5E 4	6.5E 3	3.9E-1	8.3E 3	4.2E-4	4.2E 4	1.8E 1	5.2E3
110	8.0E 3	1.5E 3	3.0E-1	6.8E 2	4.2E-4	1.0E 4	1.6E 1	6.6E3
120	4.4E 3	5.5E 2	2.3E-1	4.2E 2	3.9E-4	5.4E 3	1.3E 1	9.9E3
130	2.0E 3	2.3E 2	1.8E-1	2.3E 2	3.3E-4	2.5E 3	9.8E 0	1.4E4
150	6.5E 2	5.6E 1	1.2E-1	1.0E 2	2.6E-4	8.0E 2	6.7E 0	2.0E4
200	7.2E 1	6.5E 0	8.2E-2	2.7E-1	2.2E-4	1.0E 2	5.8E 0	3.2E4
250	1.3E 1	9.0E- 1	6.3E-2	1.0E 1	2.0E-4	2.5E 1	1.6E 1	1.2E4
300	2.9E 0	1.5E- 1	5.4E 2	4.3E 0	2.0E-4	7.4E 0	1.2E 2	1.9E3
400	1.4E-1	4.6E- 3	3.4E-2	7.1E 0	1.9E-4	9.0E-1	2.7E 2	8.8E2
500	6.9E-3	1.4E- 4	2.3E-2	1.3E-1	1.7E-4	1.6E-1	1.6E 2	1.4E3
600	3.9E-4	5.0E- 6	1.6E-2	2.5E-1	1.6E-4	4.1E-2	1.0E 2	2.5E3
700	2.2E-5	2.1E- 7	1.1E-2	5.0E-2	1.5E-4	6.3E-3	5.6E 1	4.6E3
800	1.4E-6	9.2E- 9	7.6E-3	1.0E-2	1.4E-4	1.9E-3	3.4E 1	7.8E3
900	1.0E-7	4.4E-10	5.3E-3	2.3E-2	1.3E-4	8.9E-4	1.9E 1	1.4E4
1000	7.4E-9	2.2E-11	3.7E-3	5.2E-3	1.2E-4	5.5E-4	1.2E 1	2.3E4

Z - Height

ν_{eN_2} - Collision frequency of electrons with N_2

ν_{eO_2} - Collision frequency of electrons with O_2

ν_{eHe} - Collision frequency of electrons with He

ν_{eO} - Collosion frequency of electrons with O

ν_{eH} - Collision frequency of electrons with H

ν_{em} - Collision frequency of electrons with air

ν_{ei} - Collision frequency of electrons with ions

l_{ei} - mean free path of electrons for electron-ion collisions

3.5.4. Effective frequencies of ion collosions in the Earth ionosphere [32]

	Day-time					Night-time	
Z [km]	ν_{NO+m} [Hz]	ν_{O+m} [Hz]	l_{im} [m]	ν_{ii} [Hz]	l_{ii} [m]	ν_{ii} [Hz]	l_{ii} [m]
60	6.0E 6	3.5E 6	1.0E -4	0.003	2.1E5		
70	1.8E 6	1.0E 6	3.0E -4	0.010	5.1E4		
80	2.6E 5	1.4E 5	2.0E -3	0.050	8.9E3	0.001	6.9E5
90	3.5E 4	2.0E 4	1.8E -2	0.390	1.3E3	0.004	1.5E5
100	7.3E 3	4.9E 3	7.5E -2	3.000	1.8E2	0.055	9.8E3
110	1.7E 3	9.3E 2	3.6E -1	3.200	1.9E2	0.058	1.0E4
120	6.8E 2	3.9E 2	1.0E 0	2.300	3.0E2	0.045	1.6E4
130	2.4E 2	1.4E 2	3.9E 0	1.900	4.2E2	0.035	2.2E4
150	5.8E 1	3.6E 1	1.7E 1	2.000	4.8E2	0.022	4.4E4
200	7.1E 0	5.2E 0	1.3E 2	2.500	6.8E2	0.020	6.0E4
250	2.1E 0	1.8E 0	4.3E 2	3.900	4.8E2	0.078	2.0E4
300	7.6E-1	7.5E-1	1.1E 3	5.500	3.5E2	0.670	2.3E3
400	1.5E-1	1.8E-1	1.0E 4	5.000	3.9E2	1.900	8.3E2
500	4.3E-2	5.6E-2	3.6E 4	2.700	7.7E2	1.200	1.4E3
600	1.3E-2	1.8E-2	1.2E 5	0.820	2.8E3	0.770	2.1E3
700	4.5E-3	5.8E-3	4.0E 5	0.430	5.9E3	0.400	3.9E3
800	1.8E-3	2.1E-3	1.0E 6	0.220	1.2E4	0.250	7.3E3
900	7.7E-4	7.9E-4	0.0E 0	0.200	2.0E4	0.210	1.4E4
1000	4.0E-4	3.3E-4	0.0E 0	0.160	2.9E4	0.130	2.3E4

Z - Height

ν_{NO+m} - Collision frequency of NO^{+} with air

ν_{O+m} - Collosion frequency of O^{+} with air

l_{im} - mean free path of ions for ion-air collisions

ν_{ii} - Collision frequency for ion-ion collisions

l_{ii} - mean free path of ions for ion-ion collision

3.6. Mars

3.6.1. Nominal models; daily and summer-seasonal mean midlatitude atmosphere of Mars [51]

Z [km]	T [K]	p/p_o	Northern Summer p [hPa]	Northern Summer ρ [kgm^{-3}]	g [kgm^{-3}]	Southern Summer p [hPa]	Southern Summer ρ [kgm^{-3}]
0	214.0	1.000E 0	6.36E 0	1.56E-2	3.730	7.30E 0	1.78E-2
2	213.8	8.330E-1	5.30E 0	1.30E-2	3.725	6.08E 0	1.49E-2
4	213.4	6.940E-1	4.41E 0	1.08E-2	3.720	5.07E 0	1.24E-2
6	212.4	5.790E-1	3.68E 0	9.07E-3	3.716	4.23E 0	1.04E-2
8	209.2	4.810E-1	3.06E 0	7.65E-3	3.712	3.51E 0	8.78E-3
10	205.0	3.990E-1	2.54E 0	6.47E-3	3.708	2.91E 0	7.42E-3
12	201.4	3.300E-1	2.10E 0	5.45E-3	3.703	2.41E 0	6.25E-3
14	197.8	2.715E-1	1.73E 0	4.57E-3	3.699	1.98E 0	5.24E-3
16	194.6	2.229E-1	1.42E 0	3.81E-3	3.695	1.63E 0	4.37E-3
18	191.4	1.825E-1	1.16E 0	3.17E-3	3.690	1.33E 0	3.64E-3
20	188.2	1.489E-1	9.47E-1	2.63E-3	3.686	1.09E 0	3.02E-3
22	185.2	1.211E-1	7.70E-1	2.18E-3	3.682	8.84E-1	2.50E-3
24	182.5	9.820E-2	1.79E-1	1.79E-3	3.678	7.17E-1	2.05E-3
26	180.0	7.950E-2	5.06E-1	1.47E-3	3.673	5.81E-1	1.69E-3
28	177.5	6.410E-2	4.08E-1	1.20E-3	3.669	4.68E-1	1.38E-3
30	175.0	5.160E-2	3.28E-1	9.81E-4	3.664	3.76E-1	1.13E-3
32	172.5	4.140E-2	2.63E-1	7.98E-4	3.660	3.02E-1	9.16E-4
34	170.0	3.310E-2	2.11E-1	6.48E-4	3.656	2.42E-1	7.44E-4
36	167.5	2.637E-2	1.68E-1	5.24E-4	3.652	1.93E-1	6.01E-4
38	164.8	2.090E-2	1.33E-1	4.23E-4	3.648	1.53E-1	4.86E-4
40	162.4	1.660E-2	1.06E-1	3.40E-4	3.643	1.22E-1	3.90E-4
42	160.0	1.310E-2	8.33E-2	2.72E-4	3.638	9.56E-2	3.12E-4
44	158.0	1.032E-2	6.56E-2	2.17E-4	3.634	7.53E-2	2.49E-4
46	156.0	8.100E-3	5.15E-2	1.73E-4	3.630	5.91E-2	1.99E-4
48	154.1	6.340E-3	4.03E-2	1.37E-4	3.626	4.63E-2	1.57E-4
50	152.2	4.950E-3	3.15E-2	1.08E-4	3.622	3.62E-2	1.24E-4
52	150.3	3.860E-3	2.45E-2	8.54E-5	3.618	2.81E-2	9.80E-5
54	148.7	2.990E-3	1.90E-2	6.69E-5	3.614	2.18E-2	7.68E-5
56	147.2	2.319E-3	1.47E-2	5.24E-5	3.609	1.69E-2	6.01E-5
58	145.7	1.792E-3	1.14E-2	4.09E-5	3.605	1.31E-2	4.69E-5
60	144.2	1.382E-3	8.79E-3	3.19E-5	3.601	1.01E-2	3.66E-5
62	143.0	1.063E-3	6.76E-3	2.47E-5	3.597	7.76E-3	2.84E-5
64	142.0	8.170E-4	5.20E-3	1.91E-5	3.592	5.97E-3	2.19E-5
66	141.0	6.260E-4	3.98E-3	1.48E-5	3.588	4.57E-3	1.70E-5
68	140.0	4.790E-4	3.04E-3	1.14E-5	3.584	3.49E-3	1.31E-5
70	139.5	3.670E-4	2.33E-3	8.75E-6	3.580	2.67E-3	1.00E-5
72	139.0	2.800E-4	1.78E-3	6.70E-6	3.576	2.04E-3	7.69E-6
74	139.0	2.140E-4	1.36E-3	5.12E-6	3.572	1.56E-3	5.88E-3
76	139.0	1.637E-4	1.04E-3	3.92E-6	3.568	1.19E-3	4.50E-6
78	139.0	1.252E-4	7.96E-4	3.00E-6	3.564	9.14E-4	3.44E-6
80	139.0	9.570E-5	6.09E-4	2.29E-6	3.559	6.99E-4	2.63E-6
82	139.0	7.320E-5	4.66E-4	1.75E-6	3.555	5.35E-4	2.01E-6
84	139.0	5.610E-5	3.57E-4	1.34E-6	3.551	4.10E-4	1.54E-6
86	139.0	4.290E-5	2.73E-4	1.03E-6	3.547	3.13E-4	1.18E-6
88	139.0	3.290E-5	2.09E-6	7.87E-7	3.543	2.40E-4	9.03E-7
90	139.0	2.518E-5	1.60E-4	6.03E-7	3.539	1.84E-4	6.92E-7
92	139.0	1.929E-5	1.23E-4	4.62E-7	3.535	1.41E-4	5.30E-7
94	139.0	1.479E-5	9.41E-5	3.54E-7	3.531	1.08E-4	4.06E-7
96	139.0	1.134E-5	7.21E-5	2.71E-7	3.527	8.28E-5	3.11E-7
98	139.0	8.700E-6	5.53E-5	2.08E-7	3.523	6.35E-5	2.39E-7
100	139.0	6.670E-6	4.24E-5	1.60E-7	3.519	4.87E-5	1.84E-7

p_o = 6.36hPa Northern Summer p_o = 7.3hPa Southern Summer   Z - Height

3.7. Jupiter

3.7.1. Reference model for upper atmosphere [52]

Z [km]	T [K]	P [hPa]	ϱ [kgm^{-3}]	c_s [m/s]	mean particle velocity [m/s]	coll. frequ. [Hz]
0	165.0	1.00E 3	1.61E-1	962	1256	2.03E10
10	159.3	6.89E 2	1.15E-1	949	1236	1.43E10
20	153.6	4.71E 2	8.13E-2	935	1215	9.95E 9
30	148.0	3.19E 2	5.70E-2	921	1194	6.87E 9
40	142.5	2.14E 2	3.96E-2	907	1172	4.70E 9
50	137.0	1.42E 2	2.72E-2	892	1151	3.18E 9
60	141.7	9.43E 1	1.75E-2	906	1170	2.08E 9
70	146.3	6.39E 1	1.15E-2	918	1189	1.39E 9
80	150.9	4.39E 1	7.67E-3	931	1208	9.39E 9
90	155.5	3.07E 1	5.20E-3	943	1226	6.46E 9
100	160.0	2.17E 1	3.57E-3	955	1244	4.51E 9
110	162.2	1.55E 1	2.52E-3	960	1252	3.20E 8
120	164.4	1.12E 1	1.79E-3	966	1261	2.29E 8
130	166.5	8.13E 0	1.29E-3	971	1269	1.66E 8
140	168.6	5.95E 0	9.30E-4	976	1277	1.20E 8
150	170.7	4.38E 0	6.76E-4	982	1285	8.81E 7
160	172.8	3.25E 0	4.96E-4	987	1293	6.49E 7
170	174.9	2.43E 0	3.65E-4	992	1300	4.82E 7
180	176.9	1.82E 0	2.71E-4	997	1308	3.60E 7
190	178.9	1.38E 0	2.08E-4	1002	1315	2.70E 7
200	181.0	1.05E 0	1.52E-4	1007	1323	2.04E 7
210	182.9	7.99E-1	1.15E-4	1011	1330	1.55E 7
220	184.9	6.14E-1	9.75E-5	1016	1337	1.19E 7
230	186.9	4.74E-1	6.69E-5	1020	1344	9.11E 6
240	188.8	3.68E-1	5.14E-5	1025	1351	7.04E 6
250	190.7	2.87E-1	3.97E-5	1029	1358	5.46E 6
260	192.6	2.25E-1	3.08E-5	1029	1365	4.26E 6
270	194.5	1.77E-1	2.40E-5	1026	1371	3.34E 6

Z - Height

3.7.2. Electron density in the Jovian magnetosphere measured by /5/ from [31]

date	scet	n_{max} [cm^{-3}]	n_{min} [cm^{-3}]	r/r_{Jup}
5.07.1979	3:59	3.10E-1	3.10E-1	71.4
5.07.1979	4:47	2.02E-1	2.02E-1	71.4
5.07.1979	5:35	9.34E-2	9.34E-2	71.4
5.07.1979	6:23	8.34E-2	8.34E-2	71.4
5.07.1979	58:23	4.23E-1	4.23E-1	71.0
5.07.1979	59:11	1.19E-1	1.19E-1	70.9
5.07.1979	59:59	7.11E-2	7.11E-2	70.9
5.07.1979	100:47	9.00E-2	9.00E-2	70.9
5.07.1979	2159:59	6.15E-3	6.15E-3	60.2
5.07.1979	2200:47	7.00E-3	7.00E-3	60.2
6.07.1979	1110:23	7.90E-3	7.90E-3	53.3
6.07.1979	1111:11	7.90E-3	7.90E-3	53.3
6.07.1979	1117:35	8.92E-3	8.92E-3	53.2
6.07.1979	1118:23	7.90E-3	7.90E-3	53.2
6.07.1979	1222:23	8.92E-3	8.92E-3	52.7
6.07.1979	1357:35	1.63E-2	1.63E-2	51.8
6.07.1979	1535:11	6.05E-3	6.05E-3	51.0
6.07.1979	1603:59	8.92E-3	8.92E-3	50.7
6.07.1979	1755:11	8.92E-3	8.92E-3	49.7
6.07.1979	1800:47	7.90E-3	7.90E-3	49.7
6.07.1979	1901:35	1.23E-2	1.23E-2	49.1
6.07.1979	2211:11	1.78E-2	1.78E-2	47.4
7.07.1979	0136:47	1.18E-2	1.18E-2	45.6
7.07.1979	1059:59	1.11E-2	1.11E-2	40.4
7.07.1979	1105:35	8.92E-3	8.92E-3	40.3
7.07.1979	1158:23	1.63E-2	1.63E-2	39.9
7.07.1979	1411:59	3.78E-2	3.78E-2	38.6
7.07.1979	1638:23	4.22E-2	4.22E-2	37.3
7.07.1979	1802:23	3.28E-2	3.28E-2	36.5

r - Distance

r_{Jup} - Jovian radius

n_{max} - Maximum electron density

n_{min} - Minimum electron density

scet - Spacecraft encounter time

3.7.3. Elemental abundances (relative to oxygen) of energetic heavy nuclei in the Magnetosphere

r/r_{Jup}	H	He	C	N	O	Na	S	Fe
4.9 - 5.8			<0.03 [24] /4/	<0.07 [24] /4/	1	0.03 [24] /4/	0.76 [24] /4/	
10.1-10.6			<0.13 [24] /5/	0.02 [24] /5/	1		0.62 [24] /5/	
10.6-25			0.48 [24] /5/	0.06 [24] /5/	1		0.18 [24] /5/	
25 - 40			0.38 [24] /5/	0.13 [24] /5/	1		<0.18 [24] /5/	
40 - 47	230.0 [23] /4/	3.30 [23] /4/	0.22 [23] /4/	0.13 [24] /5/	1	0.07 [23] /4/	0.34 [23] /4/	<0.02 [23] /4/
40 - 47			0.38 [24] /5/				<0.18 [24] /5/	
47 - 49	6.4 [23] /4/	0.33 [23] /4/	<0.02 [23] /4/	0.13 [24] /5/	1	0.06 [23] /4/	0.35 [23] /4/	<0.009 [23] /4/
47 - 49			:0.38 [24] /5/				<0.18 [24] /5/	
49 - 59			0.38 [24] /5/	0.13 [24] /5/	1		<0.18 [24] /5/	
59 - 70	1480.0 [23] /4/	7.40 [23] /4/	0.13 [23] /4/	0.13 [24] /5/	1	0.04 [23] /4/	0.35 [23] /4/	<0.02 [23] /4/
59 - 70			0.38 [24] /5/				<0.18 [24] /5/	

r - Distance
r_{Jup} - Jovian radius

3.7.4. Observations of Io Torus Plasma [25]

	Particle number density [cm^{-3}]								Ion temperature
r/r_{Jup}	O^+	O^{2+}	Na^+	S^+	S^{2+}	S^{3+}	SO^+	SO_2^+	T_i [K]
4.96	250	48	<20	91	14	<5.6	<2	3.5	
5.30	350	26	<72	1100	39	<3.5	<8	8	
6.00	1100	160		470	560	170		8	6E5
8.60	32	28		11	16			8	6E5

r - Distance
r_{Jup} - Jovian radius

3.8. Missions

1	Helios	11	IMP 4
2	Pioneer 10	12	HEOS 1
3	Pioneer 11	13	Prognoz 1
4	Voyager 1	14	Apollo 12
5	Voyager 2	15	Mars 2
6	ISEE 3	16	Mariner 4
7	Mariner 10	17	Mars 5
8	Venera 9	18	VEGA 1
9	Pioneer Venus	19	Giotto
10	IMP 3	20	Suisei

4. COMETARY NUCLEI

4.1. Rotation periods and obliquities of the orbit plane to the equator for some comets

Comet	designation or apparition	rotation period [h]	obliquity [degree]	reference
Bennet	1970 II	10.8	-	[1]
P/Borrelly	1912-1931	-	98 - 130	[1]
Coggia	1974 III	9.2	-	[1]
Daniel	1907 IV	16	91	[1]
P/d'Arrest	1870,1963	8.9	-	[1]
P/d'Arrest	1976	5.2	90	[1]
Donati	1958 VI	4.6	-	[1]
P/Encke	1786	-	95	[1]
P/Encke	1855	-	103	[1]
P/Encke	1901	6.5	91	[1]
P/Encke	1947	-	91	[1]
P/Encke	1977	-	84	[1]
Finsler	1937 V	7.1	-	[2]
P/Halley	1835,1910	10.3	-	[1]
P/Halley	1986	53	<15	[3]
Honda-Bernasconi	1948 IV	13.4	-	[2]
P/Kopff	1983	9.4	140	[4]
P/Schwassmann-Wachmann I	1927-1977	119	65	[1]
P/Schwassmann-Wachmann II	1930	-	35 - 50	[1]
P/Swift-Tuttle	1862	66.5	80	[1]
P/Tempel 2	1920-1967	-	65 - 75	[1]
van Gent	1941 VIII	33.6	-	[2]
Whipple-Fedtke-Tevzadze	1943 I	6.7	-	[2]

4.2. Nuclear dimensions

Comet	observation	dimensions [km]	reference
P/Halley	1986	14 * 7.5	[3]
P/Halley	1986	15 * 10	[5]
P/Kopff	1983	1.45	[4]
P/Schwassmann-Wachmann I	1983	40	[6]

4.3. Water production rates

Comet	observation	heliocentric distance [AU]	production rate $[10^{28}\ s^{-1}]$	reference
Bradfield 1979 X	Jan. 31 1980	1.03	5.1	[7]
Bradfield 1979 X	Mar. 3 1980	1.55	0.5 - 1.0	[7]
P/Encke	Oct. 24 1980	1.01	0.96	[7]
P/Halley	Mar. 9 1986	0.83	40	[8]
P/Halley	Mar. 14 1986	0.9	55	[9]
Meier 1980 q	Dec. 7 1980	1.52	8.5	[7]
Seargent 1978 XV	Oct. 16 1978	0.87	33	[7]
P/Stephan-Oterma	Dec. 7 1980	1.58	3.0	[7]
P/Tuttle	Dec. 7 1980	1.02	6.2	[7]

5. MINERALOGICAL COMPOSITION OF METEORITES

5.1. Major minerals and their chemical composition

GROUP	MAJOR MINERAL	MODE 1) [WGT-%] RANGE	MODE 1) [WGT-%] AVERAGE	MINERAL COMPOSITION [VOL-%] EN/FO/AB	MINERAL COMPOSITION [VOL-%] FS/FA/OR	MINERAL COMPOSITION [VOL-%] WO/AN 2)	MINOR- AND TRACE ELEMENTS [WGT-%]
Chondrites							
EL	ENSTATITE-GROUP	62,0 - 79,0	71,0	98 - 99	0,4-0,6		AL2O3 < 0,6; MNO < 0,01; NA2O < 0,05
	IRON - NICKEL	2,0 - 28,0	18,9				SI 1,0-2,1; CO 0,3-0,5; P < 0,2
	TROILITE	4,6 - 16,7	7,6				TI 0,2-1,0; CR 0,5-1,8; MN,ZN < 0,2
EH	CLINOENSTATITE-GROUP	58,0 - 73,0	65,0	98 - 99	0,5-1,4		AL2O3 0,3; CAO < 0,3; MNO < 0,4; CR2O3 < 0,7
	IRON - NICKEL	17,0 - 24,0	22,0				SI 2,3-3,8; CO 0,4-0,8; P < 0,7
	TROILITE	6,0 - 10,0	8,0				TI 0,2-1,0; CR 0,5-3,4; MN < 0,4; ZN < 0,2
H	OLIVINE-GROUP	25,0 - 40,0		80 - 84	16 - 20		
	ENSTATITE-GROUP	20,0 - 35,0			14 - 18		CAO 0,3-0,9
	IRON - NICKEL	14,0 - 20,0	16,8				
	PLAGIOCLASE-GROUP	5,0 - 10,0		79 - 83	5 - 8	12 - 13	FEO 0,1-0,2
	TROILITE	3,0 - 7,0	5,0				
L	OLIVINE-GROUP	35,0 - 60,0		75 - 79	21 - 25		
	ENSTATITE-GROUP	25,0 - 35,0		78 - 81	19 - 22		CAO 0,4-2,0
	IRON - NICKEL	4,0 - 12,0	6,5				
	PLAGIOCLASE-GROUP	5,0 - 10,0		83 - 87	5 - 6	9 - 11	FEO 0,1-0,2
	TROILITE	4,0 - 7,0	5,5				
LL	OLIVINE-GROUP	50,0 - 60,0		68 - 74	26 - 32		
	ENSTATITE-GROUP	20,0 - 25,0		74 - 78	22 - 26		CAO 0,4-0,9
	PLAGIOCLASE-GROUP	5,0 - 10,0		84 - 88	1 - 5	10 - 11	FEO 0,1
	TROILITE	4,0 - 6,0	5,0				

CV	OLIVINE-GROUP		70,0	86 - 94	6 - 14		
CO	OLIVINE-GROUP		70,0	77 - 91	9 - 23		
CM	PHYLLOSILICATES		70,0				
CI	PHYLLOSILICATES		63				
	EPSOMITE		7,0				
	MAGNETITE		5,0				
Achondrites							
AUB	ENSTATITE-GROUP	74,8 - 97,5	97,0	99 -100	< 0,1	< 1,2	AL2O3,MNO < 0,1; CR2O3 < 0,03; TIO2 < 0,2
DIO	ENSTATITE-GROUP		95,0	71 - 74	21 - 28	1 - 3	AL2O3 0,5-1,1; MNO 0,6; CR2O3 0,4; TIO2 0,1
URE	OLIVINE-GROUP	49,4 - 82,0		76 - 85	15 - 24		MNO 0,5; CAO 0,3-0,4; CR2O3 0,8; AL2O3 < 0,1
	PIGEONITE	15,3 - 49,1		74	18	8	AL2O3 0,4
BRA	OLIVINE-GROUP		77	67 - 70	30 - 33		CAO 0,25; MNO 0,4; P2O5,NIO 0,04; TIO2 0,01
	AUGITE		8,0	46 - 49	13 - 14	38 - 41	AL2O3 0,8-0,9; CR2O3 1,0; TIO2 0,3; NA2O 0,5
	PLAGIOCLASE-GROUP		8,0	76 - 89	2	9 - 22	
EUCO	ENSTATITE-GROUP	5,4 - 36,4	22,3	57 - 63	35 - 40	2 - 4	AL2O3,TIO2 0,2-0,5; MNO 0,8-1,0; CR2O3 0,2
	PIGEONITE	8,9 - 45,4	21,4	32 - 63	31 - 53	6 - 15	AL2O3 0,6-1,2; TIO2 0,2-0,6; MNO 0,3-0,8
	AUGITE	6,3 - 22,3	13,7	20 - 44	13 - 50	30 - 43	AL2O3 0,7-1,3; TIO2 0,3-0,6; MNO 0,3-0,8
	PLAGIOCLASE-GROUP	33,1 - 39,9	36,8			84 - 95	
EUCP	ENSTATITE-GROUP	6,2 - 38,3	18,0	52 - 64	33 - 44	3 - 4	
	PIGEONITE	10,8 - 43,6	29,8	57 - 70	23 - 37	5 - 10	
	AUGITE	4,1 - 17,8	12,1	37 - 43	17 - 21	36 - 43	
	PLAGIOCLASE-GROUP	24,8 - 39,8	33,4	5 - 34	0 - 2	65 - 94	FEO,MGO < 0,5
EUCC	ENSTATITE-GROUP	25,3 - 80,4	54,4				AL2O3 < 2,0
	PIGEONITE	0,0 - 28,8	12,4				
	AUGITE	3,7 - 15,1	6,8				
	PLAGIOCLASE-GROUP	2,1 - 60,5	23,6	5 - 8	< 0,6	91 - 95	
HOW	ENSTATITE-GROUP	37,0 - 79,7	54,7	63 - 78	21 - 33	1 - 4	AL2O3 0,5-1,0; CR2O3 0,2-0,7; MNO 0,4-0,8
	PIGEONITE	4,6 - 25,7	16,4	37 - 60	31 - 52	9 - 11	AL2O3 0,4-1,4; CR2O3 0,1-0,2; MNO 0,8

GROUP	MAJOR MINERAL	MODE 1) [WGT-%] RANGE	AVERAGE	MINERAL COMPOSITION [VOL-%] EN/FO/AB	FS/FA/OR	WO/AN 2)	MINOR- AND TRACE ELEMENTS [WGT-%]
HOW	AUGITE	3,1 - 7,2	5,4	30 - 45	9 - 34	30 - 46	AL2O3 0,5-1,0; CR2O3 0,2-0,4; MNO 0,3-0,6
	PLAGIOCLASE-GROUP	9,7 - 25,6	18,6	4 - 34	1	65 - 95	FEO 0,1-0,9
SHE	PIGEONITE	73,2 - 75,8	74,5	21 - 61	26 - 61	12 - 20	AL2O3 0,8; CR2O3 < 0,5; MNO 0,6; TIO2 < 0,4
	MASKELYNITE	17,7 - 19,1	18,4	43 - 57	1 - 3	41 - 57	FE2O3 0,6-0,7; BAO 0,01; P2O5 0,2-0,3
CHA	OLIVINE-GROUP		92,0	67	33		MNO 0,6; CAO 0,2; NIO 0,06; CR2O3 0,04
NAK	AUGITE	80 - 90		20 - 39	21 - 68	9 - 38	AL2O3 0,6-0,9; TIO2,CR2O3 < 0,4; MNO 0,4-0,6
	OLIVINE-GROUP		15	22 - 34	66 - 78		CAO 0,2-0,4; MNO 1,1; NIO,TIO2 < 0,025
ADOR	AUGITE		93,0	33	12	55	AL2O3 10,0; TIO2 2,2; CR2O3 0,2; MNO 0,06
Stony-iron-meteorites							
PAL	OLIVINE-GROUP	37,0 - 85,3	65,0	79 - 89	11 - 21		P2O5 < 5,0; MNO 0,2-0,5; CAO,NIO < 0,01
	IRON - NICKEL	14,2 - 48,6	30,3				
MES	ENSTATITE-GROUP	23,3 - 49,4		50 - 77	21 - 45	1 - 6	AL2O3 0,2-0,3; CR2O3 0,3-0,7; TIO2 0,2-0,4
	IRON - NICKEL	17,5 - 56,6					NI 6,1-10,1; GE 37-56 ppm; GA 8,9-16 ppm
	PLAGIOCLASE-GROUP	10,4 - 16,5				80 - 98	
LOD	ENSTATITE-GROUP		38,8	80 - 84	14 - 17	3	
	OLIVINE-GROUP		41,8		12 - 14		CAO 0,09
	IRON - NICKEL	NI 9,3	9,8				NI 9,3
SID	ENSTATITE-GROUP		33,0	85	14	1	AL2O3 0,2; CR2O3 0,4; MNO 0,6
	CLINOENSTATITE-GROUP			85	14	1	AL2O3 0,1; CR2O3 0,4; MNO 0,5
	IRON - NICKEL		51,0				NI 9,08; GE 0,134 ppm; GA 2,27 ppm
	TRIDYMITE		8,5				AL 2100 ppm
	TROILITE		7,0				NI 2,5; CR 2,5

Iron meteorites

I AB	IRON - NICKEL		NI 6,4-25; GE 25-520 ppm; GA 11-100 ppm
I C			
II AB	IRON - NICKEL		NI 5,3-6,4; GE 107-185 ppm; GA 46-62 ppm
II C	IRON - NICKEL		NI 9,8-11,5; GE 88-114 ppm; GA 37-39 ppm
II D	IRON - NICKEL		NI 9,8-11,3; GE 82-98 ppm; GA 70-83 ppm
II E	IRON - NICKEL		NI 7,5-9,7; GE 60-75 ppm; GA 21-28 ppm
III AB	IRON - NICKEL		NI 7,1-10,5; GE 27-47 ppm; GA 16-23 ppm
III CD	IRON - NICKEL		NI 10,0-23,0; GE 1,4-70 ppm; GA 1,5-27 ppm
III E	IRON - NICKEL		NI 8,2-9,0; GE 34-37 ppm; GA 17-19 ppm
III F	IRON - NICKEL		NI 6,8-7,8; GE 0,7-1,1 ppm; GA 6,3-7,2 ppm
IV A	IRON - NICKEL		NI 7,4-9,4; GE 0,09-0,14 ppm; GA 1,6-2,4 ppm
IV B	IRON - NICKEL		NI 16-26; GE 0,03-0,07 ppm; GA 0,17-0,27 ppm
SOROTI	IRON - NICKEL		NI 12,88; GE 5,22 ppm; GA 14,1 ppm
	TROILITE	60,0	

1) The table contains only minerals with a mode > 5 WGT-%
2) Abbreviations for the components of mixed crystals:

enstatite-group	EN- enstatite, FS - ferrosilite, WO - wollastonite
olivine-group	FO - forsterite, FA - fayalite
plagioclase-group	AB - albite, OR - orthoclase, AN - anorthite

Abbreviation for meteoritic groups

undifferentiated meteorites

EH - enstatite chondrites
EL - enstatite chondrites
H - ordinary chondrites
L - ordinary chondrites
LL - ordinary chondrites
CV - carbonaceous chondrites (Vigarano - group)
CO - carbonaceous chondrites (Ornans - group)
CM - carbonaceous chondrites (Mighei - group)
CI - carbonaceous chondrites (Ivuna - group)

differentiated meteorites

AUB - aubrites
DIO - diogenites
URE - ureilites
BRA - Brachina
EUCO - ordinary eucrites
EUCP - polymict eucrites
EUCC - cumulate eucrites
HOW - howardites
SHE - shergottites
CHA - chassignites
NAK - nakhlites
ADOR - Angra dos Reis
PAL - pallasites
MES - mesosiderites
LOD - lodranites
SID - siderophyres
I AB,...- chemical groups of iron meteorites

References:

all groups	1 - 3
EH,EL	4 - 17
H,L,LL	18 - 32
CI,CM,CO,CV	33 - 53
AUB	7,54,55
DIO	56,57,94
URE	58 - 63
BRA	64,65
EUCO,EUCP,EUCC	57,66 - 69
HOW	57,68,70 - 75
SHE,CHA,NAK	64,76 - 84
ADOR	85
PAL	64,86 - 88
MES	68,89 - 94
LOD	95
SID	96 - 98
iron groups	99 - 108
Soroti	109,110

Mineral	EL	EH	H	L	LL	CV	CO	CM	CI	AUB	DIO	URE	CHA	BRA	EUCO	EUCP	EUCC	HOW	SHE	NAK	ADOR	PAL	MES	LOD	SID	IAB	IC	IIAB	IIC	IID	IIE	IIIAB	IIICD	IIIE	IIIF	IVA	IVB	SOROTI
Element minerals																																						
COPPER	o		o	o		o		o		o		+										o	o			o		o			o	o						o
KAMACITE	#	#	#	#	+	+	+	o		+	+	+	o	o	o	o	o	o			o	#	#	#	#	#	#	#	#	#	#	#	#	#	#	#	#	#
TAENITE	+	+	+	+	+	o	o	o		o	?	+	o		?	?	?	?			?	#	#	#	#	#	#	+	#	#	#	#	#	#	#	#	#	#
TETRATAENITE			+	+	+		+				o											o	+			o						o						
AWARUITE						o																				o												
MOLYBDENUM																																						
RUTHENIUM						o																																
PLATINUM						o																																
GRAPHITE	+	+				?		o	o	o												o	o			+	+	+		o	o			o				o
DIAMOND												+														o												
LONSDALEITE																										o												
CHAOITE												o																										
SULFUR						o	o	o	+																													
Nitrides, carbides, silicides, phosphides																																						
ROALDITE																										o		o										
SINOITE	o																																					
CARLSBERGITE																											o	+				+		o				
OSBORNITE	o	o								o																												
HAXONITE																										?	?	o		o			+	+				
COHENITE	o	o										+											o			+	+	o				o	o					
SUESSITE												o																										
PERRYITE	o	o								o												o																
BARRINGERITE																						o																
SCHREIBERSITE	+	+					o	o		+		o										+	o	o	+	+	+	+	+	+	+	+	+	+	+	+	+	+

Mineral	EL	EH	H	L	LL	CV	CO	CM	CI	AUB	DIO	URE	CHA	BRA	EUCO	EUCP	EUCC	HOW	SHE	NAK	ADOR	PAL	MES	LOD	SID	IAB	IC	IIAB	IIC	IID	IIE	IIIAB	IIICD	IIIE	IIIF	IVA	IVB	SOROTI
Sulfides																																						
HEAZLEWOODITE						o																				o												
DJERFISHERITE		o								o																o						o						o
PENTLANDITE					o	+	?	+	+		o		o	+								o										o	o					o
MACKINAWITE			o	o																		o	o									o						+
TROILITE	#	#	#	#	#					+	+	+	o	+	+	+	+	+		+	+	+	+	#	#	+	+	+	+	+	+	+	+	+	+	+	+	#
PYRRHOTITE						+	+	+	+										+			o																
NININGERITE		+								o		o																										
OLDHAMITE	+	+								o		o																				o						
SPHALERITE	o	o								o													o			o	o					o						o
ALABANDITE	+	o								o													o			o						o						+
CHALKOPYRITE					o	?	?													o												o						
CASWELLSILVERITE										o																												
CUBANITE			o	o					+														o			o												?
DAUBREELITE	+	+								+		o														+	+	+	o			+		o	o	+	o	o
BREZINAITE																																						
HEIDEITE										o																												
SMYTHITE		o																																				
GREIGITE		o																																				
PYRITE						?	?						o							o																		
MOLYBDENITE						o																																
Halogenides																																						
HALITE								o				o																										
SYLVITE								o				o																										
LAWRENCITE		o				?		o	?													o	o															

Oxydes, hydroxides

SPINEL		o	o	o		+	+	+											o		+																	o
MAGNETITE			o	o	o	o	o	+	#						o			o	+	+	o	o	o															
HERCYNITE						o		o																														
CHROMITE			+	+	+	o		+			+	o	+	+	+	+	+	+	o			+	+	+	+	+	+	+	+	+	+	+	+	+	+	+	+	+
ULVOSPINEL																			o																			
TREVORITE									o																													
CaAl4O7						o																																
HIBONITE		o			o	o	o	o																														
CORUNDUM						o																																
HAEMATITE							o																															
ILMENITE			+	+	+	o					o	o	o		+	+	o	+	+	o		o	+								o		o					
PEROVSCITE		o	o	o	o	o	o	o																														
PSEUDOBROOKITE				o																																		
ARMALCOLITE						o																																
"FESON"								#																														
(Ca,Mg)Ti3O7-Ti2Ti2O7						o																																
QUARTZ	o	o								o					o	o		o		?												?	?					
TRIDYMITE	o	o								o	o				+	+	+	+		?			+		#						o	?	?			+		
CRISTOBALITE	o	o	o							o					o			o		?										o		?	?					
BADDELEYITE													o			o			o		o																	
RUTILE			o	o	o		o						o							o		o	o			o					+							
BRUCITE								o	o																													
"LIMONITE"								o	+											?																		

Silicates

OLIVINE-GROUP		o	#	#	#	#	#	#	+	+	+	#	#	#	o	o	o	+	o	#	+	#	+	#		+		o			o		o					
RINGWOODITE				o																																		
WADSLEYITE				o																																		
MONTICELLITE-GROUP			o																		o																	
GROSSULARITE						o																																
ANDRADITE						o																																
MAJORITE				o																																		
ZIRCON																o							o			o												
TITANITE																o																						
CHLORITOID							?																															
MELILITE-GROUP					o	+	+	o																														
CLINOENSTATITE-GROUP		#	+	+	+	+	?	#		+	o							+				o			#													
PIGEONITE							#		?			#			#	#	#	#	#	+			+															
PYROXFERROITE												o							o																			
DIOPSIDE	o		+	+	+	+	+	o	?	+														+		+					+							

Mineral	EL	EH	H	L	LL	CV	CO	CM	CI	AUB	DIO	URE	CHA	BRA	EUCO	EUCP	EUCC	HOW	SHE	NAK	ADOR	PAL	MES	LOD	SID	IAB	IC	IIAB	IIC	IID	IIE	IIIAB	IIICD	IIIE	IIIF	IVA	IVB	SOROTI
AUGITE		o	+	+	+	+	+	o	?		o	o	+	#	#	#	#	#	#	#	#	?	+															
UREYITE																							o			o		o										
ENSTATITE-GROUP	#	+	#	#	#	?	?	#	+	#	#		+	o	#	#	#	#				?	#	#	#	+					+		o					
WOLLASTONITE						o																																
RHOENITE						o																																
KRINOVITE																										o												
RICHTERITE		o								o																o												
KAERSUTITE													o						o	?																		
CORDIERITE						o																																
ROEDDERITE		o																								o												
YAGIITE																															o							
MERRIHUEITE				o																																		
PLAGIOCLASE-GROUP	+	+	#	#	#	+	+	o		+	+	o	+	#	#	#	#	#	?	#	o		#	+		+					+		o					
KALIFELDSPARE												o			o																o							
SANIDIN													o							+						o					o							
CELSIAN																					o																	
NEPHELINE			o	o		o																																
SODALITE			o	o		o		o																														
GLASS			+	+	+	+	+	+	+	+	+	o	+	o	o	o	o	o	+	o		o									+							
MASKELYNITE			+	+	+										+				#												o							
Phosphates																																						
GRAFTONITE																													o			o				o		
SARCOPSIDE																													o		o	o				o		
FARRINGTONITE																						+	o								o		o					
STANFIELDITE																						+	o															
MERRILLITE			+	+	+				o		?				o	o	?	o	+		+	+	+	?		o		?			+		o			?		
PANETHITE																																	o					
BRIANITE																										o							o					
BUCHWALDITE																																o						
FLUORAPATITE															o	o	?	o																				o
CHLORAPATITE			+	+	+			?			o	+						o	o	o			o			o					o							
HYDROXYAPATITE																			o																			

Sulfates,wolframates

GYPSUM						o	+
ASTRAKANITE							o
EPSOMITE, HEXAHYDRITE						+	#
SCHEELITE				o			

Carbonates,oxalates

CALCITE	o	o	o	?	?	+	o
ARAGONITE						o	
VATERITE							?
SIDERITE						o	o
BREUNNERITE						o	+
DOLOMITE							o
WHEWELLITE						o	

\# The mineral is a major constituent of the group or the individual meteorite (on an average 5 WGT-%).

\+ The mineral is widely distributed in a meteoritic group and sometimes even major constituent, but normally between 0,1 and 5 WGT-% .

o The mineral is sparsely distributed and not known from all meteorites of a group (<0,1 WGT-%).

? The occurence of this mineral is exceedingly doubtful.

6. References

References for chapter 1

[1] LANDOLT-BOERNSTEIN: 1981, 1. Edition, Vol. II, Springer, New York, 129.
[2] STONE, E. C. and LANE, A. L.: 1979, Science 204, No. 4421, 925.
[3] MORRISON, D. and SAMZ, J.: 1980, NASA-SP-439, 141.
[4] POLLACK, J. B., BURNS, J. A. and TAUBER, M. E.: 1979, Icarus 37, 587.
[5] DERMOTT, S. F.: 1979, Icarus 37, 575.
[6] BELTON, M. J. S.: 1980, Icarus 42, 71.
[7] SEIDELMANN, P. K.: 1980, Icarus 44, 19.
[8] STONE, E. C. and MINER, E. D.: 1982, Science 215, 499.
[9] SMITH, B. A. et al.: 1982, Science 215, 504.
[10] CHRISTY, J. W. and HARRINGTON, R. S.: 1980, Icarus 44, 38.
[11] ELLIOT, J. L., DUNHAM, E., WASSERMAN, L. H., MILLIS, R. L. and CHURMS, J.: 1978, Astron. J. 83, 980.
[12] STONE, E. C. and MINER, E. D.: 1986, Science 233, 39.
[13] SMITH, B. A. et al.: 1986, Science 233, 43.
[14] IAU Circ.: 1986, No. 4164.
[15] IAU Circ.: 1986, No. 4168.
[16] MASURSKY, H.: 1986, private communications.
[17] GAFFNEY: 1980, Icarus 44, 511.
[18] SZETO, A. M. K.: 1983, Icarus 55, 133.
[19] ESPOSITO, L. W., D'CALLAGHAN, M. and WEST, R. A.: 1983, Icarus 56, 439.
[20] MARSDEN, B. G.: 1982, Catalogue of cometary orbits, IAU, Cambridge.
[21] BATRAKOW, J. W., ASCHKOWA, N. W., JSWEKOW, W. A., CHANINA, F. B. and SCHOR, W. A.: 1984, Efemeridye malykh planet, Nauka, Leningrad.

Reference for chapter 2

[1] DZIEWONSKI, A. M. and ANDERSON, D. L.: 1981, PEPI 25, 297.

References for chapter 3

[1] SISCOE, G. L.: 1979, in C. F. Kennel, L. J. Lanzerotti and E. N. Parker (ed.), Solar System Plasma Physics, vol. 2, North-Holland Publishing Company, Amsterdam/New York, p. 319.
[2] DRYER, M., RIZZI, A. W. and SHEN, W.: 1973, Astroph. Sp. Sci. 22, 329.
[3] MILLET, J., LAFON, J. P. L. and LAMY, P. L.: 1980, Astron. Astroph. 92, 6.
[4] SLAVIN, J. A. and HOLZER, R. E.: 1981, JGR 86A, 11401.
[5] SLAVIN, J. A. and HOLZER, R. E.: 1979, JGR 84A, 2076.
[6] CRAVENS, T. E., NAGY, A. F., BRACE, L. H., CHEN, R. H. and KNUDSEN, W. C.: 1979, GRL 6, 341.
[7] WOLFF, R. S., STEIN, R. F. and TAYLOR, H. A.: 1982, JGR 87A, 8118.
[8] MIHALOV, J. D., SPREITER, J. R. and STAHARA, S. S.: 1982, JGR 87A, 10363.
[9] SMIRNOV, V. N., VAISBERG, O. L. and INTRILIGATOR, D. S.: 1980, JGR 85A, 7651.
[10] TATRALLYAY, M., RUSSELL, C. T., MIHALOV, J. D. and BARNES, A.: 1983, JGR 88A, 5613.
[11] AGARWAL, D. C. and PARDE, D. C.: 1981, Indian J. Theor. Phys. 29, 151.
[12] VASYLIUNAS, V. M.: 1983, in R. L. Carovillano and J. M. Forbes (ed.), Solar-Terrestrial Physics, Reidel Publishing Comp., Dordrecht, p. 479.
[13] SLAVIN, J. A. et al.: 1980, JGR 85A, 7625.
[14] BIRMINGHAM, T. J.: 1983, Rev. Geoph. Space Phys. 21, 375.
[15] NESS, N. F. et al.: 1981, Science 212, 211.
[16] SLAVIN, J. A., SMITH, E. J., GAZIS, P. R. and MIHALOV, J. D.: 1983, GRL 10, 9.
[17] WOLFF, R. S. and MENDIS, D. A.: 1983, JGR 88A, 4749.

[18] ROSENBERG, R. L., KIVELSON, M. G. and COLEMAN, P. J.: 1978, JGR 83A, 4165.
[19] BURLAGA, L. F., LEPPING, R. P., BEHANNON, K. W., KLEIN, L. W. and NEUBAUER, F. M.: 1982, JGR 87A, 4345.
[20] MULLAN, D. J.: 1983, ASTROPH. J. 272, 325.
[21] MARSCH, E., MUEHLHAEUSER, K.-H., SCHWENN, R., ROSENBAUER, H., PILIPP, W. and NEUBAUER, F. M.: 1982, JGR 87A, 52.
[22] SEIFF, A.: 1983, in D. M. Hunten, L. Colin, T. M. Donahue and V. I. Moroz (ed.), Venus, The University of Arizona Press, Tucson, p. 218.
[23] KRIMIGIS, S. M. et al.: 1979, Science 204, 998.
[24] VOGT, R. E. et al.: 1979, Science 206, 984.
[25] SULLIVAN, J. D. and SISCOE, G. L.: 1982, in D. Morrison (ed.), Satellites of Jupiter, The University of Arizona Press, Tucson, p. 309.
[26] MIHALOV, J. D. and WOLFE, J. H.: 1978, Solar Phys. 60, 399.
[27] BURLAGA, L. F., KLEIN, L. W., LEPPING, R. P. and BEHANNON, K. W.: 1984, JGR 89A, 10659.
[28] KNUDSEN, W. C. and MILLER, K. L.: 1982, JGR 87A, 2246.
[29] NESS, N. F., ACUNA, M. H. and BEHANNON, K. W.: 1982, JGR 87A, 1369.
[30] HARTLE, R. E., SITTLER, E. C., OGILVIE, K. W., SCUDDER, J. D., LAZARUS, A. J. and ATREYA, S. K.: 1982, JGR 87A, 1383.
[31] GURNETT, D. A., SCARF, F. L., KURTH, W. S., SHAW, R. R. and POYNTER, R. L.: 1981, JGR 86A, 8199.
[32] GUREVICH, A. V.: 1978, Nonlinear phenomena in the ionosphere, Springer-Verlag, New York.
[33] RUSSELL, C. T.: 1979, Phys. Earth Planet. Inter. 20, 237.
[34] WHANG, Y. C.: 1977, JGR 82, 1024. C. P.: 1963, Science 139, 909.
[35] SMITH, E. J., DAVIS, L., COLEMAN, P. J. and SONETT, C. P.: 1963, Science 139, 909.
[36] DOLGINOV, S. S., YEROSHENKO, Y. G. and ZHUGOV, L. N.: 1976, JGR 81, 3353.
[37] LANDOLT-BOERNSTEIN. Numerical data and Functional Relationships in Science and Technology. New Series. Editor in chief: K.-H. Hellwege, Group VI, vol. 2a, Editors: K. Schaefer and H. H. Voigt, Springer-Verlag, Berlin, Heidelberg, New York, 1981, p. 178.
[38] DRYER, M. and HECKMANN, G. R.: 1967, Solar Phys. 2, 112.
[39] GRINGAUZ, K. I. et al.: 1974, Kosm. Issled. 12, 585.
[40] PHILLIPS, J. L. and RUSSELL, C. T.: 1986, JGR, submitted.
[41] RUSSELL, C. T.: 1978, Proc. 9th Lun. Planet. Conf., Houston, p. 3137.
[42] SMITH, E. J., DAVIS, L. and JONES, D. E.: 1976, in T. Gehrels (ed.), Jupiter, The University of Arizona Press, Tucson, p. 783.
[43] ACUNA, M. H. and NESS, N. F.: 1976, JGR 81, 2917.
[44] COWLEY, S. H.: 1980, Nature 287, 775.
[45] SMITH, E. J. et al.: 1980, Science 207, 407.
[46] NESS, N. F., BEHANNON, K. W., LEPPING, R. P. and WHANG, Y. C.: 1976, Icarus 28, 479.
[47] HILL, T. W. and MICHEL, F. C.: 1975, Rev. Geoph. Space Phys. 13, 967.
[48] RUSSELL, C. T. et al.: 1975, Space Res. 15, 621.
[49] BUSSE, F. W.: 1976, Phys. Earth Planet. Interiors 12, 350.
[50] OGILVIE, K. W., SCUDDER, J. D., VASYLIUNAS, V. M., HARTLE, R. E. and SISCOE, G. L.: 1977, JGR 82, 1807.
[51] SEIFF, A.: 1982, Adv. Space Res. 2, no. 2, 3.
[52] PETROPOULIS, B. and BAMOS, C.: 1984, Astron. Astroph. (Suppl.) 58, 145.
[53] GRINGAUZ, K. I. et al.: 1985, Adv. Space Res. 5, no. 12, 165.
[54] JOHNSTONE, A. et al.: 1986, Nature 321, 344.
[55] NEUBAUER, F. M. et al.: 1986, Nature 321, 352.
[56] MUKAI, T., MIYAKE, W., TERESAWA, T., KITAYAMA, M. and HIRAO, K.: 1986, Nature 321, 299.
[57] NESS, N. F. et al.: 1986, Science 233, 85.

References for chapter 4

[1] MENDIS, D. A. et al.: 1985, Fundamentals of Cosmic Physics 10, 36.
[2] FERNANDEZ, J. A. and JOCKERS, K.: 1982. Rep. Progr. Phys. 46, 163.

[3] SAGDEEV, R. Z. et al.: 1986, Nature 321, 262.
[4] SEKANINA, Z.: 1984, Astron. J. 89, 1573.
[5] KELLER, H. U. et al.: 1986, Nature 321, 320.
[6] CRUIKSHANK, D. P. and BOWN, R. H.: 1983, Icarus 56, 377.
[7] FELDMAN, P. D.: 1982, in: "Comets", L. L. Wilkening (Ed.), University of Arizona Press, Tucson, p. 461.
[8] KRASNOPOLSKY, V. A. et al.: 1986, Nature 321, 269.
[9] KRANKOWSKY, D. et al.: 1986, Nature 321, 326.

References for chapter 5

[1] DODD, R. T.: 1981, Meteorites — a Petrologic — Chemical Synthesis, Cambridge University Press, Cambridge.
[2] WASSON, J. T.: 1974, Meteorites — Classification and Properties, Springer-Verlag, New York.
[3] KIESL, W.: 1979, Kosmochemie, Springer-Verlag, Wien.
[4] RUBIN, A. E.: 1983, Proc. Lunar Planet. Sci. Conf. 14th, 293.
[5] WEEKS, K. S. and SEARS, D. W. G.: 1985, Geochim. Cosmochim. Acta 49, 1525.
[6] SEARS, D. W., KALLEMEYN, G. W. and WASSON, J. T.: 1982, Geochim. Cosmochim. Acta 46, 597
[7] PETAZEV, M. I. and SKRIPNIK, A.: 1983, Meteoritika 42, 86
[8] KEIL, K.: 1968, J. Geophys. Res. 73, 6945.
[9] BHANDARI, N., SHAH, V. B. and WASSON, J. T.: 1980, Meteoritics 15, 225.
[10] FUCHS, L. H., FRONDEL, C. and KLEIN, G. Jr.: 1966, Amer. Mineral. 51, 949.
[11] EL GORESY, A. et al.: 1983, Meteoritics 18, 293.
[12] BUSECK, P. R. and HOLDSWORTH, E. F.: 1972, Meteoritics 7, 86.
[13] RAMBALDI, E. R. et al.: 1983, Meteoritics 18, 380.
[14] KEIL, K. and SNETSINGER, K. G.: 1967, Science 155, 451.
[15] ANDERSEN, C. A., KEIL, K. and MASON, B.: 1964, Science 146, 256.
[16] RUBIN, A. E.: 1984, Earth Planet. Sci. Lett. 67, 273.
[17] MCKINLEY, S. G., SCOTT, E. R. D. and KEIL, K.: 1984, Proc. Lunar Planet. Sci. Conf. 14th, P. 2, B567.
[18] BISCHOFF, A. and KEIL, K.: 1984, Nature 303, 588.
[19] HUSS, G. R., KEIL, K. and TAYLOR, G. J.: 1980, Geochim. Cosmochim. Acta 45, 33.
[20] NAGAHARA, H.: 1984, Geochim. Cosmochim. Acta 48, 2581.
[21] SCOTT, E. R. D. and TAYLOR, G. J.: 1983, Proc. Lunar Planet. Sci. Conf. 14th, 275.
[22] BISCHOFF, A. and KEIL, K.: 1983, Special Publication 22, UNM Institute of Meteoritics, Albuquerque, USA.
[23] MASON, B.: 1972, Meteoritics 7, 309.
[24] RAMBALDI, E. R.: 1981, Nature 293, 558.
[25] BUSECK, P. R. and KEIL, K.: 1966, Amer. Mineral. 51, 1506.
[26] VAN SCHMUS, W. R. and RIBBE, P. H.: 1968, Geochim. Cosmochim. Acta 32, 1327.
[27] VAN SCHMUS, W. R.: 1969, Earth-Sci. Rev. 5, 145.
[28] DODD, R. T.: 1971, Miner. Mag. 38, 451.
[29] NOONAN, A. F. et al.: 1978, Meteoritics 13, 584.
[30] FUJIMAKI, H. et al.: 1981, Mem. Nat. Inst. Polar Res. No. 20, 119.
[31] VAN SCHMUS, W. R. and RIBBE, P. H.: 1969, Geochim. Cosmochim. Acta 33, 637.
[32] SCOTT, E. R. D., LUSBY, D. and KEIL, K.: 1985, Proc. Lunar Planet. Sci. Conf. 16th, 137.
[33] NAGY, B.: 1975, Carbonaceous Meteorites, Elsevier Scientific Publishing Company, Amsterdam.
[33] KERRIDGE, J. F. et al.: 1979, Earth Planet. Sci. Lett. 43, 359.
[34] HYMAN, M. and ROWE, M. W.: 1983, Proc. Lunar Plan. Sci. Conf. 13th, P. 2, A736.
[35] GROSSMAN, L.: 1980, Meteoritics 15, 296.
[36] NORO, H. et al.: 1980, Mem. Nat. Inst. Polar Res. 17, 311.
[37] FUCHS, L. H. et al.: 1973, Smiths. Contr. Earth Sci. 10, 1.
[38] MCSWEEN, H. Jr.: 1977, Geochim. Cosmochim. Acta 41, 477.

[39] McSWEEN, H. Jr.: 1977, Geochim. Cosmochim. Acta 41, 411.
[40] KURAT, G.: Earth Planet. Sci. Lett. 9, 225.
[41] MACPHERSON, G. J. and GROSSMAN, L.: 1982, Meteoritics 17, 245.
[42] GROSSMAN, L.: 1975, Geochim. Cosmochim. Acta 39, 433.
[43] FUCHS, L. H.: 1978, Meteoritics 13, 73.
[44] HAGGERTY, S. E.: 1978, Proc. 9th Lunar Plan. Sci. Conf., 1331.
[45] TOMEOKA, P. R. and BUSECK, P. R.: 1982, Nature 299, 326.
[46] EL GORESY, A. et al.: 1979, Proc. Lunar Plan. Sci. Conf. 10th, 833.
[47] GROSSMAN, L. and STEELE, I. M.: 1976, Geochim. Cosmochim. Acta 40, 149.
[48] CHRISTOPHE MICHEL-LEVY, M. et al.: 1982, Earth Planet. Sci. Lett. 61, 13.
[49] BARBER, D. J.: 1981, Geochim. Cosmochim. Acta 45, 945.
[50] SCOTT, E. R. D. and TAYLOR, J.: 1985, Proc. Lunar Plant. Sci. Conf. 15th, 699.
[51] EL GORESY, A., PALME, H., YABUKI, H., NAGEL, K., HERRWERTH, I. and RAMDOHR, P.: 1984, Geochim. Cosmochim. Acta 48, 2283.
[52] BAR-MATTHEWS, M., HUTCHEON, J. D., MACPHERSON, G. J. and GROSSMAN, L.: 1982, Geochim. Cosmochim. Acta 48, 29.
[53] RAMDOHR, P.: 1973, The opaque minerals in stony meteorites, Elsevier Publishing Company, Amsterdam.
[54] WATTERS, R. and PRINZ, M.: 1979, Proc. Lunar Planet. Sci. Conf. 10th, 1073.
[55] OKADA, A. and KEIL, K.: 1982, Amer. Mineral. 67, 132.
[56] TAKEDA, H. and MOKI, H.: 1985, Proc. Lunar. Planet. Sci. Conf. 15th, 636.
[57] TAKEDA, H. et al.: 1976, Proc. Lunar. Planet. Sci. Conf. 7th, P. 3, 3535.
[58] BERKLEY, J. L., TAYLOR, G. J. and KEIL, K.: 1980, Geochim. Cosmochim. Acta 44, 1579.
[59] BERKLEY, J. L., TAYLOR, G. J. and KEIL, K.: 1978, J. Geophys. Res. 5, 1075.
[60] KEIL, K. and BERKLEY, J. L.: 1982, Amer. Mineral. 67, 126.
[61] RAMDOHR, P.: 1972, Meteoritics 7, 565.
[62] BERKLEY, J. L.: 1976, Geochim. Cosmochim. Acta 40, 1429.
[63] VDOVYKIN, G. P.: 1970, Space Sci. Rev. 10, 483.
[64] SMITH, J. V. et al.: 1983, Proc. Lunar Planet. Sci. Conf. 14th, P. 2, B229.
[65] NEHRU, C. E. et al.: 1983, Proc. Lunar Planet. Sci. Conf. 14th, P. 2, B237.
[66] DELANEY, J. S. and PRINZ, M.: 1984, Proc. Lunar Planet. Sci. Conf. 15th, 251.
[67] TAKEDA, H., WOODEN, J. L., MORI, H., DELANEY, J. S., PRINZ, M. and NYQUIST, L. E.: 1983, Proc. Lunar Planet. Sci. Conf. 14th, B245.
[68] DUKE, M. B. and SILVER, L. T.: 1967, Geochim. Cosmochim. Acta 31, 1637.
[69] STEELE, I. M. and SMITH, J. V.: 1976, Earth Planet. Sci. Lett. 33.
[70] WARREN, P. H.: 1985, Geochim. Cosmochim. Acta 49, 577.
[71] MIYAMOTO, M., DUKE, M. B. and McKAY, D. S.: 1984, Proc. Lunar Planet. Sci. Conf. 14th, 629.
[72] IKEDA, Y. and TAKEDA, H.: 1985, Proc. Lunar Planet. Sci. Conf. 15th, 649.
[73] DYMEK, R. F. et al.: 1976, Geochim. Cosmochim. Acta 40, 1115.
[74] DESNOYERS, C. and JEROME, D. Y.: 1977, Geochim. Cosmochim. Acta 41, 81.
[75] BUCHWALD, V. F.: 1984, Phosphate Minerals (Ed.: Nriagu, J. O. and Moore, P. B.), Springer-Verlag, Berlin—Heidelberg.
[76] STOLPER, E. and McSWEEN, H., Jr.: 1979, Geochim. Cosmochim. Acta 43, 1475.
[77] STEELE, I. M. and SMITH, J. V.: 1982, Proc. Lunar Planet. Sci. Conf. 13th, P. 1, A375.
[78] SMITH, J. V. and HERVIG, R. L.: 1979, Meteoritics 14, 121.
[79] FUCHS, L. H.: 1969, Meteorite Research (Ed. P. M. Millmann), Reidel, Dordrecht, 683.
[80] FLORAN, R. J., PRINZ, M., HLAVA, P. F., KEIL, K., NEHRU, C. E. and HINTHORNE, J. R.: 1978, Geochim. Cosmochim. Acta 42, 1213.
[81] MASON, B. et al.: 1975, Meteoritics 11, 21.
[82] BOGARD, D. D., NYQUIST, L. E. and JOHNSON, P.: 1984, Geochim. Cosmochim. Acta 48, 1723.
[83] BERKLEY, J. L., KEIL, K., PRINZ, M. and GOMES, C. B.: 1980, Proc. Lunar Planet. Conf. 11th, 1089.
[84] BOCTOR, N. Z. et al.: 1976, Earth Planet. Sci. Lett. 32, 69.
[85] PRINZ, M., KEIL, K., HLAVA, P. F., BERKLEY, J. L., GOMES, C. B. and CURVELLO, W. S.: 1977, Earth Planet. Sci. Lett. 35, 317.

[86] Buseck, P. R.: 1977, Geochim. Cosmochim. Acta 41, 711.
[87] Buseck, P. R. and Holdsworth, E. F.: 1977, Miner. Mag. 41, 91.
[88] Buseck, P. R.: 1969, Science 165, 169.
[89] Mason, B. and Jarosewich, E.: 1973, Miner. Mag. 39, 204.
[90] Powell, B. N.: 1971, Geochim. Cosmochim. Acta 35, 5.
[91] Malissa, H., Jr.: 1974, Tshermaks Min. Petr. Mitt. 21, 233.
[92] Hewins, R. H.: 1984, Proc. Lunar Planet. Sci. Conf. 15th, 289.
[93] Hewins, R. H.: 1984, Proc. Lunar Planet. Sci. Conf. 14th, 257.
[94] Yanai, K. and Kojima, K.: 1982, Meteoritics 17, 300.
[95] Doerfler, G., Hecht, G. and Ploeckinger, E.: 1965, Tschermaks Min. Petr. Mitt. 10, 413.
[96] Roesler, H. J., Blankenburg, H.-J. and Baum, H.: 1978, Chem. Erde 37, 242.
[97] Reid, A. M., Williams, R. J. and Takeda, H.: 1974, Earth Planet. Sci. Lett. 22.
[98] Buchwald, V. F.: Iron Meteorites, Univ. of California Press.
[99] Buchwald, V. F.: 1977, Phil. Trans. R. Soc. Lond. A286, 453.
[100] Bunch, T. E.: 1972, Meteoritics 7, 31.
[101] Bild, R. W.: 1977, Geochim. Cosmochim. Acta 41, 1439.
[102] Malissa, H. Jr., Wichtl, M., Kiesl, W. and Salpeter, E. W.: 1978, Richerche Spettroscopiche 3, 547.
[103] Kracher, A., Kurat, G. and Buchwald, V. F.: 1977, Geochem. J: 11, 207.
[104] Javnel, A. A.: 1982, Meteoritika 40.
[105] Fuchs, H. et al.: 1967, Geochim. Cosmochim. Acta 31, 1711.
[106] El Goresy, A.: 1965, Geochim. Cosmochim. Acta 29, 1131.
[107] Scott, E. R. D. and Bild, R. W.: 1974, Geochim. Cosmochim. Acta 38, 1374.
[108] Clarke, R. S., Jr.: 1979, Meteoritics 14, 367.
[109] Kracher, A. and Kurat, G.: 1979, Meteoritics 14, 481.
[110] Henderson, E. P. and Perry, S. H.: 1957, Proc. U.S. National Museum 107, 339.

ISBN 3-05-500426-4